Innovative Processing and Manufacturing of Advanced Ceramics and Composites II

Innovative Processing and Manufacturing of Advanced Ceramics and Composites II

Ceramic Transactions, Volume 243

A Collection of Papers Presented at the
10th Pacific Rim Conference on
Ceramic and Glass Technology
June 2–6, 2013
Coronado, California

Edited by
Tatsuki Ohji
Paolo Colombo
Makio Naito
Javier E. Garay

Volume Editor
Hua-Tay Lin

The American Ceramic Society

WILEY

Published by John Wiley & Sons, Inc., Hoboken, New Jersey.
Published simultaneously in Canada.

Library of Congress Cataloging-in-Publication Data is available.

ISBN: 978-1-118-77150-1
ISSN: 1042-1122

Printed in the United States of America.

10 9 8 7 6 5 4 3 2 1

Contents

SYNTHESIS AND PROCESSING OF MATERIALS USING ELECTRIC FIELDS/CURRENTS

Preface

With continued discoveries and innovations, the field of materials synthesis and processing remains as it has been for many decades, a vibrant and fertile area for research and development. It comes, therefore, as no surprise that every Pac Rim conference has had considerable emphasis on this topic with many symposia devoted to various aspects of this field.

This Ceramic Transactions volume represents selected papers based on presentations in four symposia during the 10th Pacific Rim Conference on Ceramic and Glass Technology, June 2–6, 2013 in Coronado, California. The symposia and their organizers are:

- **Novel, Green, and Strategic Processing and Manufacturing Technologies**
  Organizers: Tatsuki Ohji, National Institute of Advanced Industrial Science and Technology (AIST), Japan; Mrityunjay Singh, Ohio Aerospace Institute, NASA Glenn Research Center, USA; Shaoming Dong, Shanghai Institute of Ceramics, China; Jow-Lay Huang, National Cheng Kung University, Taiwan; Hai-Doo Kim, Korea Institute of Materials Science, Korea; Eugene Medvedovski, Umicore Thin Film Products, USA; Alexander Michaelis, Fraunhofer IKTS, Germany; Lalit Kumar Sharma, Central Glass & Ceramic Research Institute, India; Richard D. Sisson, Jr., Worcester Polytechnic Institute, MA, USA; Hisayuki Suematsu, Nagaoka University of Technology, Japan; Nahum Travitzky, University of Erlangen-Nuremberg, Germany
- **Polymer Derived Ceramics and Composites**
  Organizers: Paolo Colombo, University of Padova, Italy; Yigal Blum, SRI International, USA; Gian Domenico Sorarù, University of Trento, Italy; Ralf Riedel, Technical University Darmstadt, Germany; Philippe Miele, University of Montpellier 2, France; Isabel Kinski, Fraunhofer Institute for Ceramic Technologies and Systems (IKTS), Germany; Raj Bordia, University of Washington, USA; Peter Kroll, The University of Texas Arlington, USA; Yuji Iwamoto, Nagoya Institute of Technology, Japan; Dong-Pyo Kim, Pohang University of Science and Technology, Korea; Yingde Wang, National University of Defence Technology, Changsha, China

- **Advanced Powder Processing and Manufacturing Technologies**
  Organizers: Makio Naito, Joining and Welding Research Institute (JWRI), Osaka University, Japan; Junichi Tatami, Yokohama National University, Japan; Lennart Bergstroem, Stockholm University, Sweden; Yuji Hotta, National Institute of Advanced Industrial Science and Technology (AIST), Japan; C. C. Huang, Hosokawa Micron Powder Systems, USA; Norifumi Isu, LIXIL Corp., Japan; Hai-Doo Kim, Korea Institute of Machinery & Materials (KIMM), Korea; Satoshi Tanaka, Nagaoka University of Technology, Japan; Tetsuo Uchikoshi, National Institute of Materials Science (NIMS), Japan; Sujanto Widjaja, Corning Incorporated, USA; Di Zhang, Shanghai Jiao Tong University, China
- **Synthesis and Processing of Materials using Electric Fields/Currents: A Symposium Honoring Prof. Zuhair Munir**
  Organizers: Javier E. Garay, University of California, CA; Manshi Ohyanagi, Ryukoku University, Japan; Eugene A. Olevsky, San Diego State University, CA; Masao Tokita, NJS Co., Ltd., Japan

The editors wish to extend their gratitude and appreciation to all the co-organizers for their help and support, to all the authors for their cooperation and contributions, to all the participants and session chairs for their time and efforts, and to all the reviewers for their valuable comments and suggestions. Thanks are due to the staff of the meetings and publication departments of The American Ceramic Society for their invaluable assistance. We want to especially acknowledge the help of Mr. Gregory Geiger of the Society. We also acknowledge the skillful organization and leadership of Dr. Hua-Tay Lin, PACRIM 10 Program Chair.

We hope that this issue will serve as a useful resource for the researchers and technologists in the field of processing and manufacturing of advanced ceramics and composites.

TATSUKI OHJI
PAOLO COLOMBO
MAKIO NAITO
JAVIER E. GARAY

Novel, Green, and Strategic Processing and Manufacturing Technologies

OPTIMIZED SHAPING PROCESS FOR TRANSPARENT SPINEL CERAMIC

Alfred Kaiser[1], Thomas Hutzler[2], Andreas Krell[2], and Robert Kremer[3]
[1]LAEIS GmbH, Wecker, Luxembourg
[2]Fraunhofer Institute for Ceramic Technologies and Systems (IKTS), Dresden, Germany
[3]ALPHA CERAMICS GmbH, Aachen, Germany

ABSTRACT

Bulky transparent ceramic, especially spinel ($MgAl_2O_4$), can be used for applications such as high-energy laser windows and lightweight armor. One of the traditional routes to manufacture transparent spinel plates includes the steps of material preparation, uniaxial pressing, cold isostatic pressing (CIP), de-bindering and hot isostatic pressing (HIP). When larger sizes are required, CIP can become one of the bottlenecks of the process chain. The paper shows, how an optimized uniaxial hydraulic pressing process with an evacuated mould allows to avoid the cold isostatic pressing completely. The process is described and the first results of the investigation of transparent spinel properties are discussed. This simplified process will allow to reduce the manufacturing costs for larger sized transparent spinel significantly and/or improve the production capacity.

INTRODUCTION

Spinels are a group of minerals of general formulation AB_2O_4 (where A is a bivalent and B is a trivalent cation), which crystallise in the cubic crystal system with A and B occupying some or all of the octahedral and tetrahedral sites in the lattice. Spinel ($MgAl_2O_4$), after which the spinel group is named, in its pure form (single crystal) is a colourless, transparent material with high hardness and excellent transmission from the ultraviolet (0.2 µm) to the mid-infrared (5 µm) region. This makes spinel an interesting material for numerous applications such as high-energy laser windows and lightweight armor[1]. However, single crystal spinel is difficult to make in dimensions greater than a few millimeters using traditional high temperature (>2000 °C) melt growth techniques[2]. Various approaches to transparent polycrystalline spinel have been made, e.g. by using sintering aids like LiF or using sub-µm spinel powder[3-8] and applying different shaping and sintering technologies[9, 10]. One of the traditional routes to manufacture polycrystalline transparent spinel plates includes the steps of material preparation, uniaxial pre-pressing, cold isostatic pressing (CIP), de-bindering and hot isostatic pressing (HIP). When larger sizes are required, CIP can become one of the bottlenecks of the process chain.

Aim of this work was to investigate a more economic route to prepare large-sized transparent spinel plates by optimization of the uniaxial pressing process and eliminating the cold isostatic pressing process completely.

MATERIAL PREPARATION

Commercially available high purity spinel powder, especially developed for transparent ceramics applications, was prepared by IKTS for the tests (see Table I).

Table I. Technical data of spinel powder used as starting material.

supplier	BAIKOWSKI, France
product code	S30CR
BET specific surface area	30 ± 5 m²/g
d_{50} (PSD Sedigraph)	0.2 µm
crystalline phase (XRD)	≥ 99 % spinel

The aggregated raw powder (Fig. 1 left) with high specific surface area (Table I) and a majority of particles ranging between 60 and 90 nm (Fig.1 right), was deagglomerated by milling in a NETZSCH horizontal disk mill type LME 4 for one hour with deionized water and with about 7 % of organic binders.

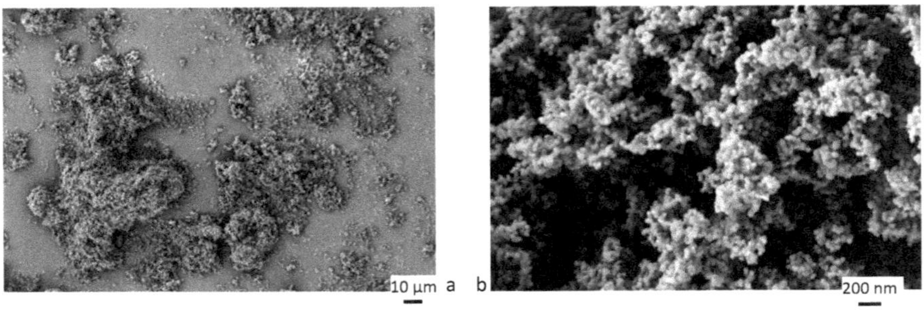

Figure 1. SEM micrographs of S30CR powder as supplied

After milling the slurry was dried in a freeze dryer type EPSILON 2-45D (supplier: CHRIST, Germany). For obtaining ready-to-press granulates the very fluffy dried spinel powder was screened with a mesh size of 250 µm. Figure 2 shows a SEM photograph of the powder as it was used for the pressing trials. Despite their flaky structure the agglomerates show a sufficient flowability for even mould cavity filling.

Figure 2. SEM micrograph of processed S30CR powder ready to press

UNIAXIAL HYDRAULIC PRESSING

For the uniaxial pressing trials a LAEIS hydraulic press type ALPHA 1500/120 was used (Figure 3), which is available at the LAEIS technical center ALPHA CERAMICS in Aachen. Features of this type of uniaxial hydraulic presses, available sizes and application examples for shaping of advanced ceramic products are described elsewhere[11-13].

Figure 3. Hydraulic press LAEIS ALPHA 1500-120

The main technical data are summarized in Table II. The press was equipped with a vacuum unit to allow the plates being pressed in an evacuated mould. Thus, a major part of the air is removed before compaction, which helps to avoid delamination or other pressing defects due to entrapped air[14].

Table II. Technical data of LAEIS hydraulic press ALPHA 1500/120.

press type		**ALPHA 1500/120**
pressing force	kN / t	15,000 / approx. 1,500
useful die area	mm x mm	1,320 x 640
maximum filling depth	mm	120
maximum ejection force	kN	300

The first tests were performed in a mould with the dimensions 191 x 142 mm^2. The mould was manually filled with a filling depth of 25 mm and pressing under vacuum took place at pressures between 100 and 250 MPa. The resulting green plates showed high strength with sharp contours and could be handled very easily. A green density of up to 1.9 g/cm^3 was reached.

Depending on the applied pressure, a densification factor between 3.5 and 4.0 was obtained. Figure 4 shows that with higher pressures the density could still be increased significantly, however the mould design capacity was limited to 250 MPa. Some of the plates were repressed in a cold isostatic press at 350 MPa, ending up with a green density of approximately 1.97 g/cm³, which is similar to what can be extrapolated from the uniaxial pressing values.

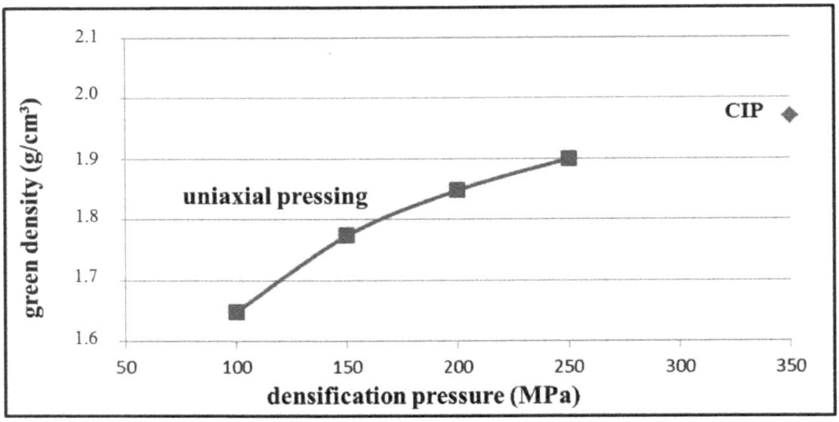

Figure 4. Green density of spinel plates 191 x 142 mm² pressed with uniaxial press (□) and with cold isostatic press (◊) at various pressures

A second test series was made on the ALPHA 1500/120 press using a mould with the dimensions 300 x 400 mm². Again the mould was filled manually with a filling depth of 50 mm and pressed under vacuum (see Figure 5). Due to the large pressing area the maximum applicable pressure in this series was 120 MPa. The green density of the large plates pressed with 120 MPa was between 1.70 and 1.71 g/cm³, which fits perfectly with the values obtained for the smaller plates (fig. 4). The densification factor was approximately 3.5 (i.e. green plate thickness of approximately 14 mm), again matching the corresponding results of the smaller plates. Also these large plates showed good strength and well defined contours and could be handled and shipped safely.

Figure 5. Volumetrically filled mould 300 x 400 mm² and green plate after ejection

These tests show that scaling up works and that it is possible to press also large-sized spinel plates without defects to a sufficient green density. If a larger press is used (presses are available up to a pressing force of 64,000 kN = 6,400 t) and the mould is designed accordingly, pressures up to approximately 350 or 400 MPa can be realized. Therefore it would be possible to press plates e.g. with green dimensions of 400 x 400 mm² to similar green densities which can be achieved by cold isostatic pressing.

THERMAL PROCESSING

All activities regarding thermal treatment of the green plates were conducted at the IKTS in Dresden. Removal of organic components (debindering) was performed at 800 °C in air. To avoid thermally induced stress, a low heating rate of 6 K/h was used. After debindering, the smaller samples were pre-sintered without pressure up to closed porosity (> 95%...98% relative density) at temperatures between 1550 and 1590 °C for 2 hours, again with low heating and cooling rates. The necessary temperature for a sufficient sintering process (elimination of the open porosity) was approx. 35 K higher for the uniaxially pressed plates compared to the CIP repressed plates, apparently due to differences in the density and in the homogeneity of particle coordination in these differently shaped bodies.

The pre-sintered plates were hot isostatic pressed in argon at 1590 °C for 15 hours and at a maximum pressure of 185 MPa. The HIP-process was performed in a hot isostatic press with graphite heaters and a useful volume of 32 dm³ (supplier: EPSI, Belgium). After HIP, all plates were at the same density of 3.58 g/cm³ and showed transparency. Solely the plates which were pressed with only 100 MPa had some opaque areas where the porosity was not eliminated completely. Figure 6 shows some photographs of samples prepared from the smaller plates after polishing.

Figure 6. Transparent spinel samples prepared from 191 x 142 mm² plates (thickness approx. 4 mm)

The large plates were pre-sintered at 1600 °C for 2 hours and hot isostatic pressed at 1600 °C for 15 hours. Plates with a density of 3.581 g/cm³ were obtained, showing good transparency except very few small areas close to the surface (see Figure 7). The overall linear shrinkage in length and width amounts to approximately 23 %, the shrinkage in thickness is even higher (27 %).

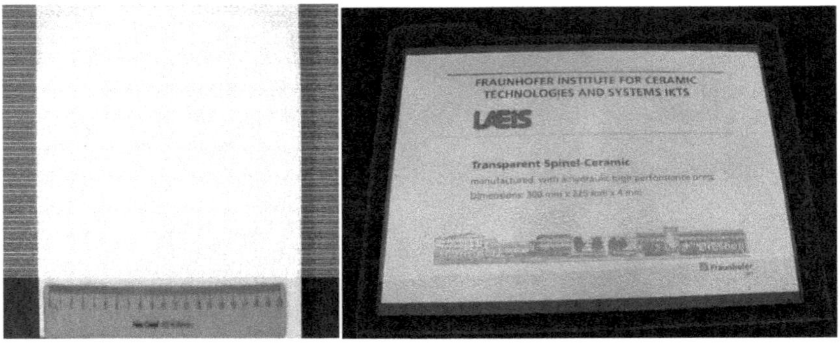

Figure 7. Large spinel plate after sintering (left) and transparent after HIP + both sides ground and polished (right)

SAMPLE CHARACTERIZATION

Samples with approximately 20 mm diameter and 4 mm thickness were prepared from the smaller plates as well as from the large plates. They were ground and polished and the real-inline-transmission (RIT) was determined. The real in-line transmission was measured at 640 nm wave length with an aperture of about 0.5° (LCRT 2000, Gigahertz Optics, Puchheim, Germany). This measurement excludes scattered amounts which are included in "in-line" data of commercial spectrometers with larger effective apertures of 3-5° [15,16]. The results are shown in Table III.

Table III. Transmission values of different spinel samples.

plate size (after HIP)	pressure		RIT	thickness
mm x mm	MPa		%	mm
110 x 130	100		78.8	3.94
110 x 130	150		79.3	3.94
110 x 130	150	+ 350 (CIP)	81.6	3.94
110 x 130	200		80.3	3.58
110 x 130	200	+ 350 (CIP)	81.4	3.94
110 x 130	250		81.4	3.87
220 x 300	120		76.1	3.94
220 x 300 *)	120		61.5 – 72.8 (69.6 – 77.1) **)	5.95

*) toll polishing in an industrial plant; surface roughness Rz = 0.17 μm compared to Rz = 0.03 μm of the laboratory prepared samples

**) recalculated to thickness = 4.0 mm

The RIT values show a comparable high fluctuation, when measured in a regular grid over the whole area of the plate. There is also a number of macroscopic visible defects (pores, contaminations?) spread over the whole plate area, and the large-sized plates have different shades of greying despite identical production conditions. The Vickers hardness of the large plate was determined to 1277 ± 24 HV10 as can be expected[17].

Average grain sizes of the dense microstructures were determined on SEM micrographs by the linear intercept approach (average grain size =1.56 average intercept length[18]). SEM mircrographs of polished cross sections show a bimodal structure (Fig. 8): an average grain size of 2.5 μm was measured in the fine grained areas whereas grain sizes between about 25 and 110 μm were observed in the coarser parts of the microstructure.

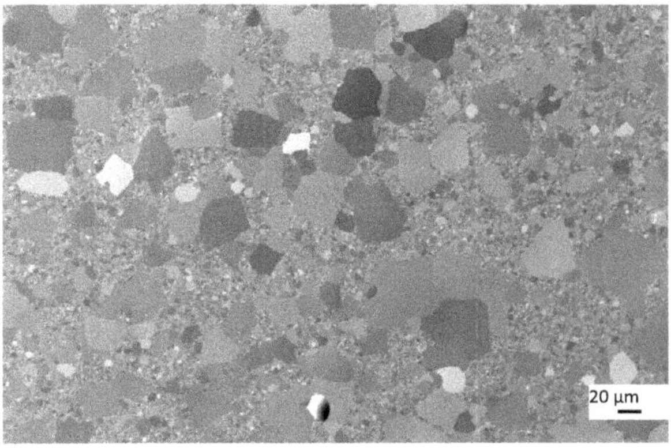

Figure 8. Example of microstructure after HIP at 1600 °C

SUMMARY AND OUTLOOK

The results reported in this paper are based on a preliminary feasibility study. It basically proved the possibility to produce crack free transparent spinel plates, also of larger size, using vacuum assisted uniaxial hydraulic pressing technology and waiving the expensive redensification by CIP. There are still open questions remaining and additional development work need to be done, eg.:

- evaluation of a powder preparation process which is better suited for large scale production than the lab scale freeze drying process
- determination of the reason(s) for the remaining visible defects and elimination of such reasons
- increasing the pressure applied during the uniaxial shaping (adaption of mould design and/or use of larger press)
- optimization of pressing parameters, vacuum application
- optimization of debindering, sintering, and HIP parameters

The intermediate results, derived from a very small number of pressing trials and sample characterization, provide a very good starting point for further evaluation and for the development of a successful production scale technology.

REFERENCES

[1]D.C. Harris; History of development of polycrystalline optical spinel in the U.S., *in: Window and Dome Technologies and Materials IX, Proceedings of SPIE Volume 5786, ed. Randal W. Tustison (2005)*
[2]J.S. Sanghera, G. Villalobos, W. Kim, S. Bayya, and I.D. Aggarwal; Transparent spinel ceramic, *NRL Review (2009)*
[3] A. Goldstein; Correlation between $MgAl_2O_4$-spinel structure, processing factors and functional properties of transparent parts, *J. Eur. Ceram. Soc. 32 (2012) [11] 2869-2886*
[4] R. Cook, M. Kochis, I. Reimanis, H-J. Kleebe; A new powder production route for transparent spinel windows: powder synthesis and window properties, *in:Window and Dome Technologies and Materials IX, Proceedings of SPIE Volume 5786, ed. Randal W. Tustison (2005)*
[5] G. Villalobos, J.S. Sanghera, and I.D. Aggarwal; Transparent ceramics: magnesium aluminate spinel, *NRL Review (2005)*
[6]I. E. Reimanis, K. Rozenburg, H-J. Kleebe, R. L. Cook; Fabrication of transparent spinel: the role of impurities, *in:Window and Dome Technologies and Materials IX, Proceedings of SPIE Volume 5786, ed. Randal W. Tustison (2005)*
[7]A. Krell, T. Hutzler, J. Klimke, A. Potthoff; Nano-processing for larger fine-grained windows of transparent spinel, *Proceedings of 34th International Conference on Advanced Ceramics and Composites; Daytona Beach, Jan 24-29, 2010, The American Ceramic Society; ed. J.J. Swab, S. Mathurs & T. Ohji, The American Ceramic Society; ed. M. Halbig & S. Mathur (= Ceram. Eng. Sci. Proc. 31 (2010)[5] 167-182).*
[8]A. Krell, T. Hutzler, J. Klimke, A. Potthoff; Fine-grained transparent spinel windows by the processing of different nanopowders, *J. Am. Ceram. Soc. 93 (2010) [9] 2656-2666*
[9]A. LaRoche, K. Rozenburg, J. Voyles, L. Fehrenbacher, G. Gilde; An economic comparison of hot pressing vs. pressureless sintering for transparent spinel armor, *in: Advances in Ceramic Armor IV: Ceramic Engineering and Science Proceedings 29 (2009) [6] (ed. L. P. Franks), John Wiley & Sons, Inc., Hoboken, NJ, USA*
[10]K. Morita, B.-N. Kim, K. Hiraga, H. Yoshida; Fabrication of high-strength transparent $MgAl_2O_4$ spinel polycrystals by optimizing spark-plasma-sintering conditions, *Journal of Materials Research 24 (2009) [9] 2863-2872*
[11]A. Kaiser, R. Lutz; Uniaxial hydraulic pressing as shaping technology for advanced ceramic products of larger size, *Interceram 60 (2011) [3] 230-234*
[12]R. Lutz; Use of closed loop controls in hydraulic press forming of ceramic products to obtain highest dimensional accuracy, *Proceedings of the International Colloquium on Refractories, Aachen (2004) 222-224*
[13]A. Kaiser; Shaping of large-sized sputtering targets, *Proceedings of 36th International Conference on Advanced Ceramics and Composites; Daytona Beach, Jan 22-27, 2012 The American Ceramic Society; ed. M. Halbig & S. Mathur*
[14]A. Kaiser, R. Kremer; Fast acting vacuum device: guaranteed quality for pressed refractories, *Interceram Refractories Manual (2003) 28-33*
[15]R. Apetz, M.P.B. van Bruggen; Transparent alumina: a light scattering model, *J. Am. Ceram. Soc. 86 (2003) [3] 480-486*
[16]H. Yamamoto, T. Mitsuoko, S. Iio; Translucent polycrystalline ceramic and method for making same, *Europ. Patent Application EP - 1 053 983 A2, IPK7 C04B35/115, 22.11 (2000)*
[17]A. Krell, A. Bales; Grain size-dependent hardness of transparent magnesium aluminate spinel, *Int. J. Appl. Ceram. Technol. 8 (2011) [5] 1108–1114*
[18]M. I. Mendelson; Average grain size in polycrystalline ceramics, *J. Am.Ceram. Soc., 52 (1969) [8] 443–6*

This paper has originally been published in the proceedings of the 37th International Conference and Exposition on Advanced Ceramics and Composites (ICACC 13), January 27 - February 1, 2013 at Daytona Beach:

A. Kaiser, T. Hutzler, A. Krell, R. Kremer; Optimized shaping process for transparent spinel ceramic, Ceramic Engineering and Science Proceedings Volume 34, Issue 8, **Advanced Processing and Manufacturing Technologies for Structural and Multifunctional Materials VII,** *Tatsuki Ohji and Mrityunjay Singh, Editors; Soshu Kirihara and Sujanto Widjaja, Volume Editors; The American Ceramic Society, 2013*

THERMAL DIFFUSION COATINGS FOR WEAR-RESISTANT COMPONENTS FOR OIL AND GAS INDUSTRY

E. Medvedovski, F.A. Chinski, J. Stewart
Endurance Technologies Inc.
Calgary, AB, Canada

ABSTRACT

Engineering components using in oil and gas industry, e.g. tubing systems for down-hole applications, slurry pump components and some other large size components made of steels and alloys, require protection against slide abrasion, impact and corrosion. CVD coatings consisted of the borides obtained through thermal diffusion of boron into the metallic base of the components provide significant improvement of the properties of the surfaces designated for the contacts with abrasive and corrosive media and enhance the integrity and service life of the components in several times. Thermal diffusion process and quality of the coatings are defined by the batch composition providing the boron deposition and diffusion, temperature and time of the process, type of the metallic base and its surface quality, engineering design of the CVD process system. The obtained coatings with thicknesses of 50-250 up to 300-400 μm (depending on the application) are uniform through the whole length of the components, do not have delamination issues, and they are successfully used in industry.

INTRODUCTION

Structural components used in mining, mineral and oil & gas processing equipment require protection against wear and corrosion due to their serious destruction in actual processing conditions. The wear and corrosion problems take place with piping systems, elbows and reducers, cyclone and hydrocyclone sections, pumping devices, centrifuges, heat exchanger tubing and many other engineering components used for transporting, separation, extracting, classifying, conveying and dust collection equipment at the processing of various hard and abrasive particles, which may be in dry and slurry forms, corrosive suspensions and emulsions. The losses due to the damage of the processing equipment components because of wear and corrosion problems and due to the associated unscheduled shutdowns rich over 17 billion dollars. However, ~40-50% of these losses can be avoided though the proper prevention methods.

Oil and gas development and production is one of the largest industrial manufacturing areas where numerous processing units and equipment components have to be protected against serious wear and corrosion problems. For example, long production tubing strings and down-hole tools with structural wear and corrosion resistant components have to be used in every oil well with utilization of the artificial lift system (e.g. sucker rod pumps or progressive cavity screw pumps). As another example, steam assisted gravity drainage (SAGD) and cyclic steam stimulation (CSS) processes, which are used for steam injection for extracting heavy oil applied for thermal stimulation of oil reservoirs and which are considered as the most efficient in oil production, require components with superior erosion, cavitation and corrosion resistance (including in water steam of high pressures and temperatures) and high mechanical strength. The materials for equipment components and tubing with high wear and corrosion resistance are strongly required for the recovery and transporting of crude oil with oilsands, i.e. oil contained abrasive mineral particles of sand and ores, which are processed in huge volumes in Canada. In addition, the sulphur recovery processes at the natural gas processing plants, e.g. the Claus process focused on the recovery of elemental sulphur and the Wet Sulphuric Acid process focused on the sulphuric acid processing, require the structural materials, particularly for piping systems, with high corrosion resistance (e.g. to sulphuric acid and sulphur-based substances).

Specific issues with tubing related to oil & gas processing (e.g. production and refinery) are dealt with the points that the tubing are mostly made from steels and alloys and have big lengths and relatively small diameters with necessity to protect the inner surfaces. The down-hole tubing is often bent at the installation due to not only mechanical issues but also due to special well designs, which required the bent tubes. Continual increasing of the well depth (up to 7,000 m that becomes a reality in the present oil development) results in the increase of the process pressures and temperatures up to 70 MPa and 250-300°C, respectively that, accordingly, requires high mechanical and structural integrity of the materials used for tubing.

Although monolithic components made of advanced ceramics and composites can be used successfully for the above mentioned application conditions[1-8], in many cases their application is limited. This is related to the needs of large and long-size components, e.g. piping systems, where the length of pipes achieves 10 m or more, or some other large size complex shape products. In this case, the surface engineering, particularly, the advanced coatings on the components (e.g. piping systems) made of steels and alloys have to be used. Different coating methods are used in industry; they include painting, dipping, cold and thermal spraying, chemical and electrochemical methods (anodizing, electroplating, electroless plating), physical vapor deposition (PVD) and chemical vapor deposition (CVD) and some others[9-14]. However, the majority of these methods have serious limitation due to the difficulties to coat long size (e.g. inner surface of small diameter tubing with several meters length) and complex shape components or due to spalling and/or delamination of the coating layer or fast destruction of thin coating layers with low hardness, i.e. lower performance, under severe wear, corrosion, temperature conditions or microcracking due to the substrate and coating thermal expansion mismatch that also results in the fast degradation of products under severe conditions. However, some surface engineering methods based on the CVD principles may be employed to overcome these serious application/ service and manufacturing challenges.

A thermal diffusion coating method that is based on the CVD principles may be successfully used for the processing of the protective coatings for the mentioned applications and for large-size components. This method involves the deposition of the selected element(s) in an active vapor form onto the preheated substrate with consequent chemical modification of the substrate metal surface through the formation of new hard inorganic substances[13]. This method allows effective protection of the entire surface of the metallic components. Among the options of thermal diffusion methods, boronizing of steels and alloys can be used to create hard and chemically inert coatings based on iron borides and some other borides, which can successfully protect metallic components from wear and corrosion at elevated temperatures[15-19]. As known, metal boride materials have a strong advantage due to their high hardness and chemical inertness and related application properties, e.g. wear and corrosion resistance[20]. For example, the iron boride coatings demonstrated lower erosion rates in comparison with the electroless nickel coatings and the coatings based on Cr_3C_2, CrB, NiCrBC, TiN, WC and some others obtained by PVD, plasma spray, nitriding and CVD methods[14, 19]. The boride coatings also demonstrated resistance to the action of corrosive environments, such as acids and salts, as well as to high-temperature oxidation[16, 18, 21-23].

Numerous publications related to the boride coating formation on steels and alloys mostly describe the results obtained in the laboratory conditions using small samples (e.g. tablets) as the substrate material. In all these publications, the processing features associated with actual high-volume production conditions are not considered. The present paper contains the results of the development, implementation and manufacturing of the boride coatings (e.g. iron boride) obtained through the thermal diffusion process on steels at Endurance Technologies Inc. (ETI). The coatings are formed on the working surface of the components, which have to serve at the extreme conditions, such as wear and corrosion, in oil and gas industry. In particular, these coating are applied for the inner surface of the tubular goods with relatively small diameters and big lengths, which are generally difficult to apply for using other methods. The obtained results, including extensive manufacturing experience, demonstrate successful applicability and high potential of the boronized coatings for severe industrial conditions.

BASIC PRINCIPLES OF THERMAL DIFFUSION BORONIZING PROCESS

The thermal diffusion method involves chemical modification of the substrate metal through the deposition of the gaseous (vapor) phase of the required atoms formed during heat treatment, its diffusion into the structure of the heated substrate material and formation of new hard inorganic phases based on the metal from the substrate and this acting atom, and, finally, diffusion and growth of these new phases, which modify the structure and properties of the base material surface. Diffusion and reaction of the ingredients formed in the gaseous state occur in significantly lower temperatures comparing with the "conventional" solid state processes. The coating formation occurs through the diffusion of the active deposited atoms into the substrate material and opposite diffusion of the atoms from the substrate driven by chemical potential gradient, resulting in the new phase formation and the growth of the coating phase.

Thermal diffusion coatings, as the material surface engineering, have advantages dealt with their high adhesion strength since strong bonds between phases of the coating and the substrate material. Since the process is diffusion controlled, a strongly visible interface between the coating and the substrate does not exist as opposed to many other coating techniques, such as PVD, electrochemical methods, painting and others; a diffusion gradient in the coating occurs instead with a formation, in some cases, of the transition zone.

With regards to the thermal diffusion boronizing process, diffusion of B into the Fe base can occur through two major mechanisms (Fig. 1): 1) diffusion through the vacancies occurs since the base steel material has surface structural defects, particularly developing at elevated temperature, and 2) interstitial diffusion occurs because the boron atomic radius is significantly smaller than the iron atomic radius (the calculated atomic radii of B and Fe are ~ 0.85-0.87 pm and ~ 1.40-1.56 pm, respectively). Diffusion of B into the lattice structure of steels results in the slight distortion of the lattice that promotes the hardness similarly to the action of carbon in irons and steels. The presence of the alloying elements in steels retards the diffusion of B and the boride layer growth. A new phase formation in the Fe-B system occurs in accordance with the phase diagram (Fig. 2)[17, 18, 24, 25]. The solubility limit of B in Fe is very small, below 0.008 wt.-% at 900°C (Fig. 2b), i.e. the presence of B in the zone outer the diffusion zone is hardly detected.

Diffusion of B into Fe can be expressed in accordance with the formula that can be written similarly to the expression described diffusion of some other intermetallic compounds[26]:

$$D = (N_{Fe}D_B + N_BD_{Fe})F_TF_{VW};$$

where N_{Fe} and N_B denote the alloy composition in mole fractions of Fe and B, D_B and D_{Fe} – diffusivity of B and Fe, F_T – thermodynamic factor, F_{VW} – vacancy-wind factor (1 – 1.5).

$$D_{B(Fe)} = a^2V;$$

where a – atomic diameter, V – frequency of atoms movement.

The formation of the diffusion layer can be expressed using the formula:

$$d\delta/dt = kSD/\delta \text{ or, after the integration, } \delta = (2kSDt)^{0.5};$$

where δ – thickness of the surface diffusion layer, S – contact area, t – time.

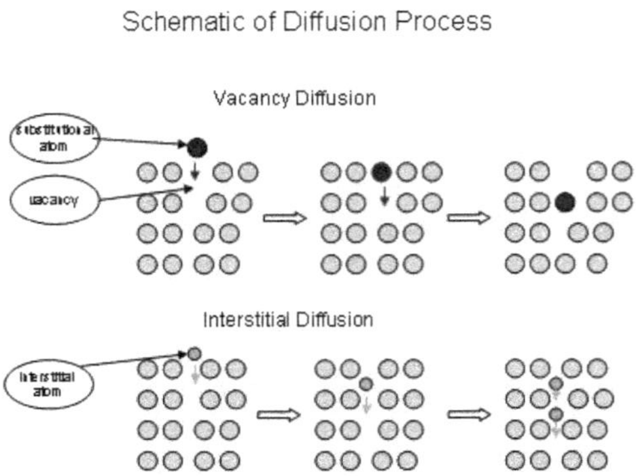

Fig. 1. Schematic of the Diffusion Routes

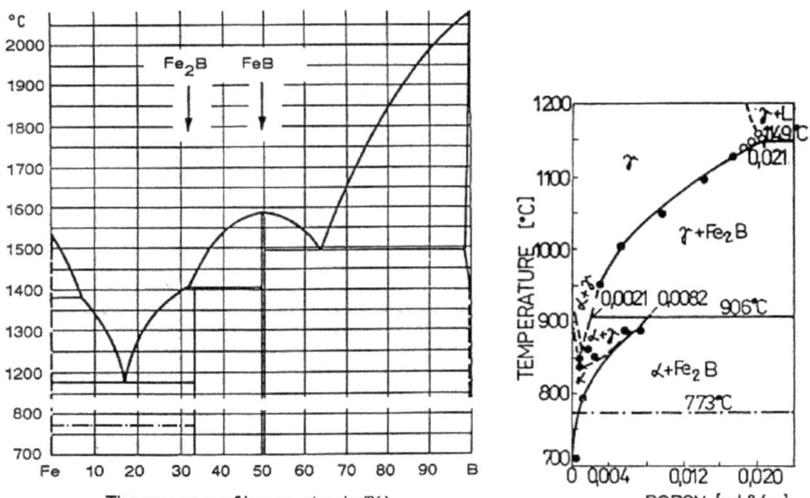

a) b)

Fig. 2. Phase Diagram and Solubility Limit in the Fe-B System
a) Phase Diagram; b) Solubility Limit

As noted above, the diffusion process is defined by temperature and time, i.e.

$$D = D_0 exp(-E/RT) = a^2 V exp(-E/RT) ;$$

where D - diffusion coefficient, T – temperature.

The formation of the iron boride thermal diffusion coatings has a parabolic behavior that is confirmed by numerous studies[14-19, 21, 27-33]. Because of this, the phase formation and, particularly, case depth (coating thickness), can be managed by the modification of temperature and time and other parameters affecting diffusion, which will be considered below.

Basically, in accordance to the Fe-B phase diagram, two phases, Fe_2B with tetragonal structure and FeB with orthorhombic structure, can be formed[16-19, 21, 24, 25]. It is considered that the FeB phase is undesirable, despite its higher hardness, since it is usually associated with higher coefficient of thermal expansion and internal tensile stresses, which may result in the micro-cracking and flaking-off[17-19]. However, in practice, both phases occur, especially when the coatings with substantial thicknesses (100 – 250 μm) are produced. In the case of the presence of alloying elements in steels, the borides of these alloying elements may form [16-20, 34], i.e. FeB, Fe_2B, (Fe,Cr)B, $(Fe,Cr)_2B$, (Fe,Ni)B, $(Fe,Ni)_2B$ and some other borides can be found.

Different B-contained materials can be used as the source of boron ("donors"). The publications mentioned the use of amorphous boron, boron-based alloys, boron carbide and some other B-contained powders, which may be taken from small to large percentage (up to 80%) in the mix [17-19, 21, 27, 28, 32-34]. The heat treatment process can be conducted in the inert conditions, e.g. in the controlled atmosphere, preventing the oxidation of boron, iron and other ingredients of the processing system. However, some of the boronizing mix compositions and process conditions mentioned in literature may not be well suitable for the industrial manufacturing.

EXPERIMENTAL

Starting Materials and Processing

Batch (pack) compositions used for the thermal diffusion boronizing process consist of three major ingredients, such as a boron-sourcing material, a process activator (initiator) and an inert filler; all ingredients are powders. As the major boron-sourcing material, boron carbide powders commercially produced were selected. The activator used for the boronizing process provides the formation of the volatile substances, which initiate the transport of the B-substance to the substrate surface and its reaction with iron. As the activator, a halide salt is used. The inert filler "restricts" an excessive interaction of the B-species with iron, promotes the atmospheric conditions preventing oxidation of boron carbide and boron, as well as it provides the conditions, when the used mix can be removed and transported from the retort easily. The ratios of the ingredients of the boronizing batch compositions, e.g. B-sourcing material : activator : inert filler are basically in the range of (1.5 – 15) : (3 – 10) : (80 – 95); however, the types of the ingredients and their particular ratios, are the proprietary of ETI. The material after the boronizing cycle consists of mostly inert powder; however, it contains some quantities of the remaining boron. Because of this, it can be utilized for the next boronizing process cycles as a filler with adequate recalculation. The pack composition is selected in accordance to the required thickness of the coating, demanding properties, as well as in accordance to the substrate material (type of steel or alloy).

The selected powder ingredients are mixed, and this homogeneous mix is used to fill a containment vessel or retort with the steel or alloy components, which have to be boronized. In some cases, depending on the required area, which has to be treated, the actual product may be as a retort. Although different steels and alloys are boronized at ETI, carbon steels with chemical compositions (wt.-%) of C 0.15-0.20, Si 0.20-0.23, Mn 0.60-0.75, Cu 0.30-0.35,Ni 0.10-0.15, Cr 0.10-0.16,Mo 0.02-0.05, and Fe as a balance are mainly used in industry as an inexpensive base (substrate) material to be boronized. The surface of the base material (substrate), which has to be coated, is preliminary cleaned and blasted using the hard powders with certain grits to remove the foreign impurities and to create a smooth working surface. The boronizing is conducted in the high temperature furnaces at the temperature range of 800 – 1100°C; the temperature, heating-cooling profile and heat treatment conditions are selected based on the requirements to the product (e.g. coating thickness and properties, etc.). The heating and cooling ramps are generally rather fast, that is dictated by the mechanical requirements to steel components, but they are also defined by the capability of the heating equipment. Because the iron boride formation is defined by the diffusion process, i.e. it depends, in a high extent, on temperature and time, these parameters can be varied to obtain the desirable case depth. The boronized products are cleaned up from the remaining spent mix and inspected. For the quality control, the witness samples are placed into the retort with the actual products, and these witness samples pass the same processing cycle as the actual products. The flow chart of the boronizing process used at ETI is shown on Fig 3. The spent mix can be reused (recycled) for the consequent boronizing cycles. The ETI process allows manufacturing of the products of different shapes and sizes, including with complex shapes and the tubing with length up to 12 m (with inner and outer coated surfaces) with homogeneous coating depth and structure.

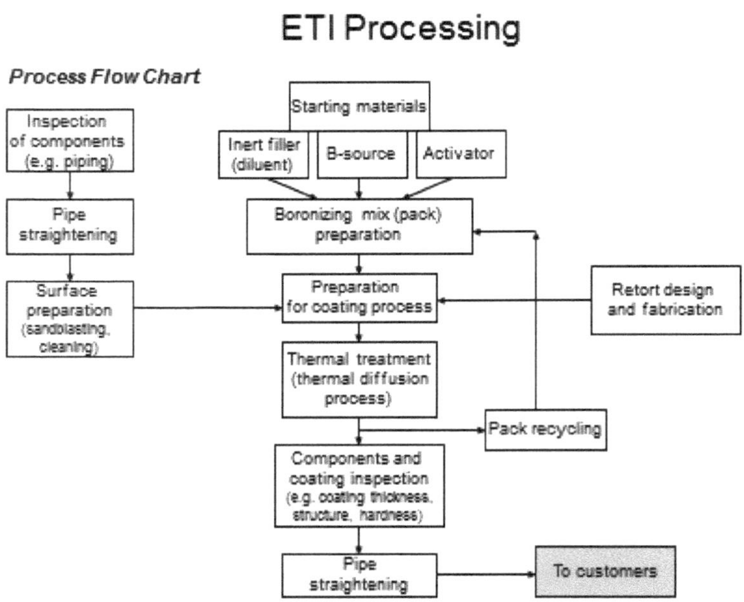

Fig. 3. Boronizing Process Flow Chart

Testing and Evaluation

The process features occurred with the batch composition during heat treatment were studied by thermogravimetric analysis (TGA) using TA Instruments Q600 Thermogravimetric Analyser. The TGA heating process was conducted in Ar to avoid the oxidation of the ingredients and related thermal effects. The heating rate was 3°C/min. and maximal temperature of the TGA run was 1200°C that exceeds the actual temperature of the boronizing process. The coating thickness (case depth), structure and morphology were determined using an optical microscope and scanning electron microscope (SEM) JEOL JSM-6610. The metallographic samples were prepared through cutting, grinding and polishing using the Buehler equipment. The relative contents of the key elements at the coating were determined by the X-ray energy dispersive spectrum (EDS) analysis using Oxford Instruments INCA x-ACT EDS detector with INCA Suite 4.13 software; the quantities both in weight-% and atomic-% were determined.

The coatings were also characterized by the Knoop hardness determination at 100 g load (HK0.1) in accordance with ASTM E384-10. The samples for testing were ground and polished using the Buehler equipment. The Knoop hardness was calculated in accordance with formula:

$$HK = 14.229xP/d^2,$$

where P – force (kgf) and d – length of the long diagonal of the replica (mm) made at the indentation.

Tensile properties of the tubular samples, in particular, Yield Strength (*YS*), ultimate tensile strength (*TS*) and % of elongation, were determined in accordance with ASTM A370-12A. The samples (~305 mm or 1 ft. of length) with the inner surface coating, as well as the uncoated samples were tested. *YS* was determined from the Stress-Strain Diagram by the 0.2% off-set method. Tensile strength values were calculated in accordance with the formula:

$$TS = R/A,$$

where R - maximal load, A - cross-sectional area.

Wear resistance of the boronized steel samples, as well as bare steel (for comparison), was determined in accordance with dry sand rubber wheel test ASTM G65, procedure B. In this test, the abrasive medium (50-70 mesh silica sand) was continuously fed between the sample and the rotating rubber lined wheel. The testing schematic is shown on Fig 4. The wear test duration was 2000 revolutions. This test is widely used for the wear resistance evaluation of different materials in the mining, mineral and oil processing.

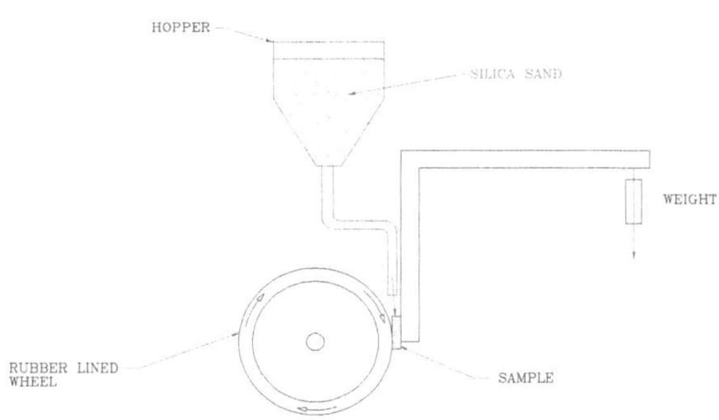

Fig. 4. Schematic of the Wear Test – ASTM G65 Dry Sand Rubble Wheel Test (Procedure B – 2000 revolutions)

Corrosion resistance was studied at elevated pressures and temperatures in the corrosive environments of water steam with the presence of hydrocarbons, $CO+CO_2$, H_2S, chloride salts ($NaCl$ and $CaCl_2$), which simulated the actual oil well corrosive conditions, using the autoclave system. In particular, testing temperature and pressure were in the range of 200-300°C and 14-28 MPa, respectively. The samples were partially immersed into the corrosive liquid using special holders, so this setting allowed to study the corrosion resistance in liquid and gaseous phases. The principle schematic of the corrosion testing unit is shown on Fig 5. The unit consists of the autoclave connected with the gas supplying vessels, compressor and pressure gage, the vent system and the pressure and temperature regulator. The corrosion test duration was 170 – 240 hrs. The coatings after the exposure in the corrosive environment were evaluated using optical microscopy and SEM.

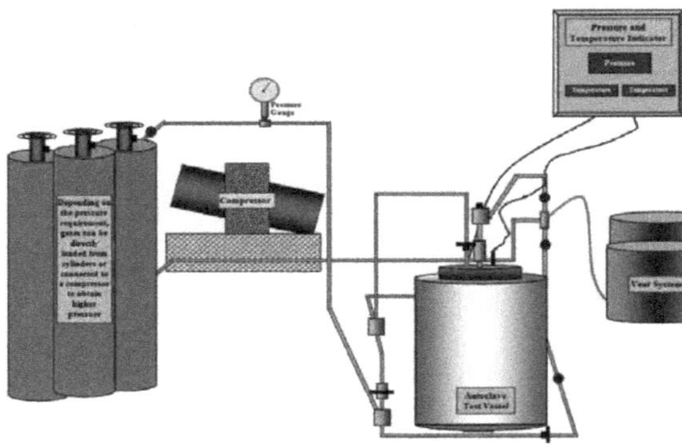

Fig. 5. Principal Schematic of the Corrosion Test Autoclave Unit

RESULTS AND DISCUSSION

The boronizing process occurs in several steps depending on the process temperature, and some of the boronizing process steps are defined by the processes occurring in the boronizing mix. They include the vapor formation and the initiation of the reaction between this vapor and the B-source ingredients and the further formation of the volatile B species. The vapor occurs from the high temperature transformation of the halide-based activator. This process is accompanied by some weight loss that can be observed from the TGA curve (Fig. 6). The effects on the TGA curve and a total weight loss depend on the composition of the boronizing mix.

The formed B-based vapor interacts with a metallic substrate, and this interaction continues with formation of iron borides and their diffusion into the substrate. Formed iron boride Fe_2B reacts with the diffusing B yielding FeB, and FeB diffuses into the substrate and interacts with Fe atoms, which also diffuse from the substrate towards to FeB with the formation of Fe_2B in accordance with the Fe-B phase diagram (Fig. 2). I.e. diffusion and interaction occurs in two directions:

$$B + Fe_2B \rightarrow 2FeB \text{ and } FeB + Fe \rightarrow Fe_2B$$

The boride layer formation starts from the nucleation, and the nuclei grow and merge through the "necking" forming a thin layer. The layer of Fe_2B grows towards to the substrate due to its tetragonal morphology. Further B diffusion into Fe-B with the formation of Fe_2B and FeB is accompanied by the coalescence of the forming grains with the filling of the structural pores and defects of the substrate material resulting in the practically dense sintered Fe-B layer.

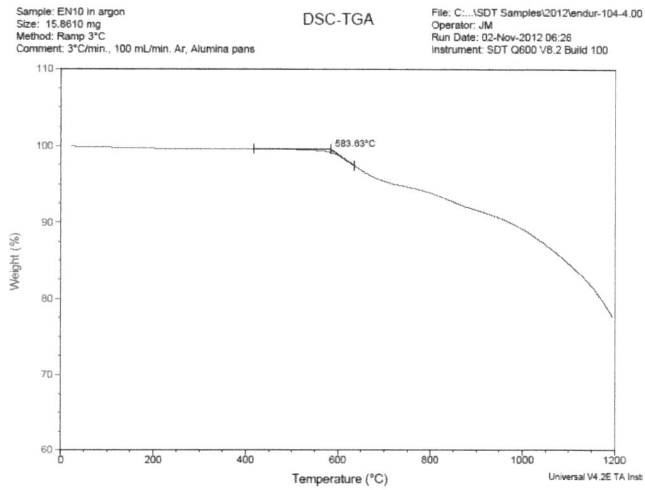

Fig. 6. TGA Curve of the Boronizing Mix

The boronizing process, as defined by diffusion, can be managed by the modification of temperature and time. From our practical experience, the process temperature and, especially, time can be varied depending on the size of the products, which have to be boronized. For example, larger components usually require larger volumes of the powder mix that surrounds these components, and the amount of the powder should be enough for the thermal diffusion to reach the required case depth. In this case, a longer soak (time) has to be used. The adjustment of time is more preferable for the industrial manufacturing, particularly for large-size products, than the adjustment of temperature; it allows to maintain better the uniformity of the temperature distribution in the production kilns and to avoid overheating of the steel products and therefore to minimize the degradation of steels at high temperatures.

As an example, an influence of the boronizing time on the case depth formation is shown on Fig. 7-8. The increase of the soak time from 6 to 15 hrs resulted in the case depth growth from ~100 to ~250 μm at the boronizing of the tiles with the areas of ~150x200 mm. Boronizing of smaller samples (e.g. 25-50x75-100 mm) requires ~1.5 times shorter soaks to reach similar case depth.

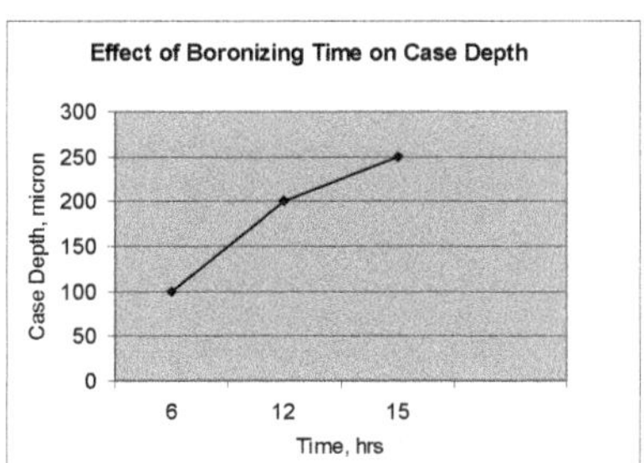

Fig. 7. Influence of Processing Time on the Case Depth

The surface preparation of the substrate also affects the diffusion process and the case depth formation. The surface treatment using physical and chemical cleaning and blasting with abrasive powders or beads provides not only the removal of grease and scale but, in a high extent, enhances the diffusion process through vacancies because the amount of crystallographic defects, such as dislocations, in the surface and subsurface zones. The substrate surface blasting promotes the case depth for about 25-50 μm in the case of the coating of carbon steels comparing with non-blasted materials. This treatment is important for the coating of stainless steels because the latter have denser structures.

The boronized coating structure usually has the "saw-tooth" morphology (Fig. 8). With regards to the case depth determination, it was measured from the top of the coating surface to the middle of the "teeth". Generally, the inner zone (which is closer to the substrate) mostly consists of the Fe_2B phase, and the outer zone mostly consists of the FeB phase. The calculations based on the data obtained from the EDS analysis confirm this point. Although some authors [28, 35, 36] claimed the formation of the single Fe_2B phase (which is mostly mentioned when very thin, below than 50 μm, coatings are obtained), in practice, when large size components are processed in large furnaces and when the soak time at the boronizing is rather long, two phases FeB and Fe_2B are formed. However, by the modification of the pack composition and process parameters, it is possible to reach a rather high yield of the Fe_2B phase. Due to optimization of the pack composition and process parameters, the Fe_2B phase and the FeB phase have similar "saw-tooth" morphology despite their different crystalline lattice parameters (see Fig. 8). I.e. the FeB phase grows on the Fe_2B repeating the morphology of the Fe_2B phase. This structure provides good bonding between the "layers" with no delamination and cracking when the columnar grains of the layers reinforce the coating in general, and this effect provides a high level of integrity of the coating and its high properties. No mechanical interface between the boride coating and the substrate is observed. The thermal diffusion process provides rather uniform coating for the entire treated surface with no delamination issues. This level of the coating structure is reached, for example, through the full length of the tubing with the length of greater than 10 m when a high level of temperature uniformity in the production furnaces is maintained.

The distribution of the major elements in different zones of the boronized coating can be seen at the SEM images with the EDS data (Fig. 9). The EDS studies were conducted for three different coating zones – the "top" zone with a thickness of only a few microns, and two zones, i.e. "intermediate" and "bottom" zones, of the actual iron boride coating. In accordance with the EDS data, the "top" zone contains a high percentage of B with some amounts of C, Fe and O; the "intermediate" and "bottom" zones have lower contents of B and higher contents of Fe that is dealt with the diffusion of B to the depth of the base material and the coating formation. The contents of C and O in these inner zones are very small. The presence of O can be explained by the slight oxidation of the ingredients during boronizing, and it is clear that the outer area has a higher level of oxidation. The presence of C is not clear understood; some level of C may be dealt with the samples preparation and the presence of C in the instrument chamber. However, it is clear that the C content is higher at the "top" zone. The comparison of the contents of B and C in the "top" zone and in B_4C (stoichiometric composition) allows to assume that the "top" area consists of the substance with a composition of $B_{4+x}C$ (i.e. boron-rich B_4C). It can be assumed that, during the vapor phase formation, not only B but also C occur in the gaseous phase, and both elements deposit onto the substrate and diffuse with the formation of the B-rich B_4C phase. However, the formation of C and its diffusion is significantly smaller than B. Also, the formation of iron carboborides $Fe_3(C,B)$ and $Fe_{2,3}(C,B)_6$ is possible [19, 34] because some amounts of Fe were detected in the "top" zone. Based on the amounts of B and Fe in the "intermediate" and "bottom" zones of the coating, it is assumed that the "intermediate" zone consists of mostly FeB phase and the "bottom" zone consists of mostly Fe_2B phase that is in a good agreement with general principles of diffusion and with phase formation at the boronizing process considered above.

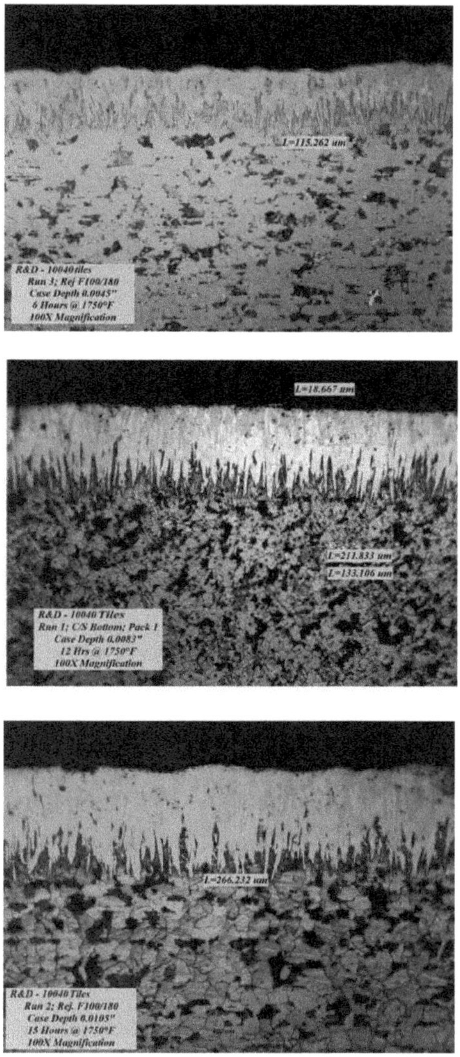

a)

b)

c)

Fig. 8. Microstructure of the Boronized Coatings at Different Process Time (optical microscope, 100^x magnification)
a) 6 hrs; b) 12 hrs; c) 15 hrs

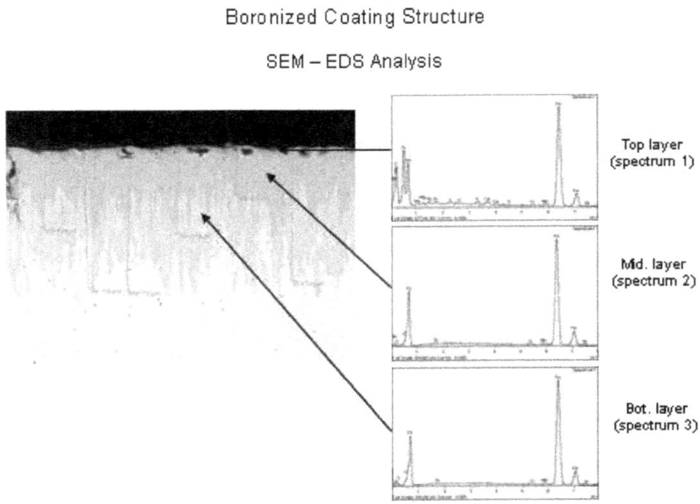

Fig. 9. Boronized Coating Structure – SEM-EDS Analysis

Based on the conducted studies and extensive manufacturing and application experience, the following factors related to the processing have a significant influence on the thermal diffusion process, case depth and structure formation and properties of the boronizing coatings:

- Type of steel or alloy, which include their chemical composition, carbon content, alloying elements presenting in steels and alloys; for example, a denser structure and a presence of the alloying elements as Cr, Ni and some others in stainless steels delay the diffusion of B. The substrate surface treatment prior to the thermal diffusion process also affects the case depth formation
- Processing features, mainly temperature, timing and gaseous environment (e.g. the processing conditions should provide the avoidance of the oxidation of boron and other elements involving in the process), as well as the uniformity of temperature distribution in production furnaces
- Pack composition (boronizing powder mix) including the ratio of ingredients, type of the B-source, powders particle size distribution, morphology of starting ingredients, pack homogeneity; these features of the pack composition define the ability to form a vapor phase and the "reactivity" of the materials involved in the process. It is important to use an optimal particle size distribution of the powder mix with a good flow to maintain uniform feeding of the retorts and to discharge the retorts and products after the boronizing run, particularly when large size components are produced
- Vapor transport that, in the significant extent, depends on the amounts of the formed vapor, vapor pressure and the particle size distribution in the pack
- Size and shape of metallic substrate component, which define the amount of the pack material required for the process and dictate the selection of temperature and time

- Retort design that has to be done based on the size and shape of the components and that should provide uniform boronizing process

These factors have to be carefully considered not only for the boronizing process but also for other processes based on the thermal diffusion approach.

The physical and mechanical properties of the boronized coatings and the steel products with these coatings are shown in Table. The Fe-B coatings of 100-250 μm thickness have hardness in about 10 times higher than uncoated carbon steels. The obtained results correlate well with other published data related to hardness of the iron boride coatings [16-19, 21, 37]. It can be specified that the inner zone of the coatings that mostly consists of the Fe_2B phase has slightly lower hardness values than the outer area of the coatings mostly consisted of the FeB phase (HK0.1 1400-1500 and 1650-1800, respectively). If the coating thickness is below 100 μm, the hardness values drop to ~1600-1650 because these thinner coatings have higher contents of the Fe_2B phase. The tensile properties of the boronized steels satisfy the ASTM requirements; however, they are at the bottom level of the specification due to the loss of ductility during the high-temperature treatment and due to the formation of the hard but lower ductile layer. The properties of the iron boride coatings obtained using all "fresh" ingredients or using the processed recycling mixes for the coating formation have the same level.

Table. Mechanical Properties of the Boronized and Uncoated Steels

Property	Testing method	Untreated steel*	Boronized steel* (100-250 μm layer)
Knoop hardness HK0.1	ASTM E384-10	160 – 180	1500 – 1800
Yield strength, 0.2% offset, MPa	ASTM A370-12A	380 – 550	380 – 410
Ultimate tensile strength, MPa	ASTM A370-12A	520 min.	600 – 660
% of elongation	ASTM A370-12A	18 – 22	18 – 20

*Carbon steel J55

Sliding abrasion testing conducted in the "dry" conditions (in accordance to ASTM G65, procedure B) demonstrated that the boronized steel performed about 10 times better than bare carbon steel (Fig. 10). The values of the wear loss (volume loss) of the tested materials were calculated from the weight loss and density. The obtained results confirmed the data of higher wear resistance of the iron boride coatings over untreated steels reported by other authors, which tested wear resistance by other methods [16-19, 35, 36, 38-40].

Our studies of the worn surfaces of the materials after the sliding abrasion test show that the wear mechanisms for the bare steel and boronized steel are different. Ductile materials, including the materials with rather low ductility (1-2% elongation) such as the carbon steel used for the testing, are worn due to the localized plastic flow occurred under the impact of hard particles and friction, which exceed the critical strain to failure in local areas, primarily the areas with surface defects. This failure occurs mostly due to the removal of the "platelets" formed under the plastic flow and "ploughing". As opposed to bare steel, wear of the hard ceramic boronized coatings is dealt with the cracking and chipping mechanism. Similarly to many other hard but brittle ceramic materials [1, 3-7, 14], wear of the boronized coatings with the hardness that is several times higher than the hardness of processing particles starts from impact of abrasive particles and micro-crack formation. Conical cracks are formed in a high extent. However, in the reality, the processing particles have both rather round and angular shapes, and their velocity may vary with high turbulence when impact and sliding occur; in this case, plastic flows, radial and lateral cracks formed. Cracks usually start in the weakest areas of the coatings, e.g. from micro-cracks dealt with the processing irregularities. Then the "network" from the cracks generates during continuation of the abrasion action. Finally, destruction at the weaker grain-boundary phase and pulling-out of the grains result in the material removal, as well as further crack generation. The ductile substrate material, as the backing to the hard coating, reduces the crack propagation (especially, radial cracks) and therefore delays wear. The "saw-tooth" structure adherent to the boronized coating and the absence of the clear mechanical interface between the coating and the substrate also promote wear resistance. Generally, the coatings with larger case depth are more preferable for the wear applications; however, if the case depth is greater than ~250-275 μm, more micro-cracks may form in the coating structure (it may be dealt with general brittleness of Fe-B coatings and with stress formation between the FeB and Fe_2B phases with different crystalline lattice structures and different coefficients of thermal expansion), which affect negatively wear-related crack formation and wear loss. In addition to significantly higher hardness and to the surface structure features, boronized coatings provide lubricity in adhesive wear applications, despite their higher roughness (comparing with untreated steels).

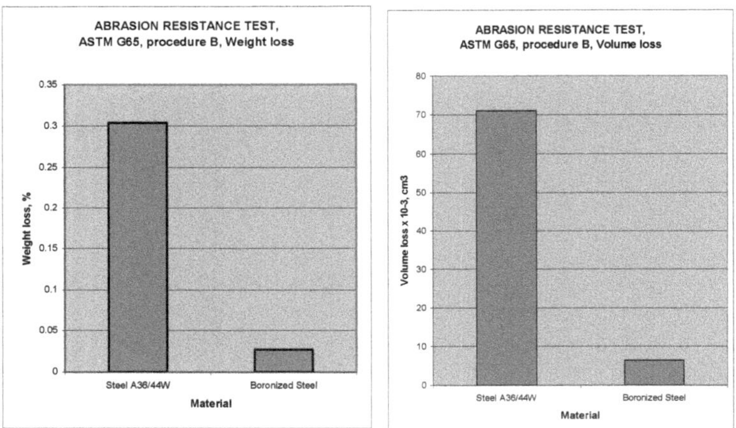

a) b)

Fig. 10. Wear Resistance Test Results (ASTM G65, Procedure B) – Weight (a) and Volume (b) Loss

The testing of the boronized samples in the highly corrosive conditions (steam with the presence of H_2S, hydrocarbons, $CO+CO_2$, chlorine salts) at elevated pressures and temperatures (up to 28 MPa and 300°C, respectively), which simulate the oil processing corrosion situations, demonstrated no structural degradation of the coatings after the 7 – 10 days exposure, i.e. no phase transformation, no cracking, spalling and delamination were observed. In comparison, not only carbon steels, but stainless steels have visible corrosion after the first hours of the corrosion testing. This testing also demonstrated good oxidation resistance of the boronized coatings because water steam is a strong oxidant, especially at elevated pressures and temperatures. The piping systems with the boronized coatings of the thickness of ~80-120 μm are successfully used at the actual oil field corrosion and corrosion-abrasion conditions in the processing of water-steam-oil-mineral mud environments with the presence of H_2S, $CO+CO_2$, S-based substances, hydrocarbons, chlorides at elevated pressures and temperatures. This thickness is selected, based on the conducted experiments, to minimize possible occurrence of the micro-cracks due to large size piping handling. The corrosion resistance of the studied coatings is defined by high thermodynamic stability, a high level of the covalent Fe-B bonds and dense structure.

ETI successfully produces tubing and casing with the boronized and some other protective coatings for the North American oil and gas industry. For example, over 80,000 joints of boronized tubing put in service over last four years; they include pup joints (EndurAlloy™ Wear Joints), sucker rods used for progressive cavity and bottom-hole rod pumping systems and other piping systems with up to 12 m length of a single component, as well as long helical components, for lifting and transporting of heavy oil. Also different slurry pump components, ESP stages, chokes, valves and fittings, spool pieces, coal/oil burner nozzles and a variety of tooling produced are served in wear and corrosion environment in oil & gas, mineral processing and refinery. Some products produced at ETI can be seen on Fig. 11. The boronized tubing and other products provide the significant extension of service life (3-10 times comparing with untreated steels depending on the application) in severe conditions and the sufficient decrease of processing losses and unscheduled shutdown of the equipment. In the case of failure, the damaged coating may be refurbished instead of the complete components replacement.

Fig. 11. Some Products with Protective Boronized Coating

CONCLUSION

- Hard iron boride coatings are produced by ETI in Calgary for large-size complex shape and long tubular components, which are used successfully in a variety of oil & gas processing and refinery applications.
- The components with the ETI coatings (EndurAlloy^{TM} Wear Joints and others) provide significant extension of service life (3-10 times comparing with untreated steels) in severe wear and corrosion environments of oil well production conditions. This provides a significant reduction of processing losses and unscheduled shutdowns.
- Versatile technology based on "all fresh" ingredients or based on used (recycling) mixes provides identical processing and coating performance. The utilization of the processed mixes provides a significant reduction of materials loss and waste and pollution minimization.
- Process-structure-properties relationship is outlined. The obtained coating structure provides a high level of application properties, as wear and corrosion resistance. Modification of thermal diffusion process parameters allows managing the case depth and coating structure for different sizes of components and for different application conditions.

ACKNOWLEDGEMENT

ETI acknowledges the financial support of the NRC Canada for the process development and optimization (IRAP project 801309). The assistance of GR Petrology (Calgary, Canada) in the EDS analysis and Charter Coating (Calgary, Canada) in the corrosion testing is appreciated greatly.

REFERENCES

[1]I.M. Hutchings, Tribology. Friction and Wear of Engineering Materials, Butterworth-Heinemann, London, UK, 1992
[2]G. Arthur (Editor), Wear Resistant Surfaces in Engineering, UK Dept. of Trade and Industry, HMSO, London, UK. 1985
[3]D.H. Buckley, K. Mioshi, Fundamental Tribological Properties of Ceramics; *Cer. Eng. Sci. Proc.,* **6**, N. 7/8, 919-939 (1985)
[4]E. Medvedovski, Wear-Resistant Engineering Ceramics; *Wear*, **249**, N. 9, 821-828 (2001)
[5]K. Kato, K. Adachi, Wear of Advanced Ceramics; *Wear*, **253**, N. 11-12, 1097-1104 (2002)
[6]B. Basu, M. Kalin, Tribology of Ceramics and Composites. A Materials Science Perspective, Wiley, Hoboken, NJ, USA, 2011
[7]E. Medvedovski, D. D. Marchant, Wear- and Corrosion-Resistant Ceramics for Protection of Pipelines and Rotating Equipment, p.223-232 in *"Proceedings of the 2000 International Pipeline Conference (IPC 2000)"*, vol.1, ASME 2000, Calgary, Alberta, Canada, October 1-5, 2000
[8]E. Medvedovski, Influence of Corrosion and Mechanical Loads on Advanced Ceramic Components, *Ceramics Int.*, **39**, N. 3, 2723-2741 (2013)
[9]P.R. Roberge, Handbook of Corrosion Engineering, McGraw-Hill, 2000
[10]K.G. Budinski, Overview of Surface Engineering and Wear, ASTM Int. Publishing, 1996, 4-21
[11]K.G. Budinski, Surface Engineering for Wear Resistance, Prentice-Hall Inc., 1988
[12]Surface Hardening of Steels: Understanding the Basics, 2002 ASM International
[13]J.R. Davis (Editor), Surface Engineering for Corrosion and Wear Resistance; ASM International and IOM Communications, Maney Publishing, 2001
[14]A. V. Levy, The Erosion-Corrosion Behavior of Protective Coatings; *Surf. Coat. Technol.*, **36**, N. 1-2, 387-406 (1988)
[15]L.G. Voroshnin, L.S. Lyakhovich, Boronizing of Steel, Metallurgy, Moscow, 1978 (in Russian)

[16]A.K. Sinha, Boriding (Boronizing); ASM Handbook. Heat Treating, **4**, 437-447 (1991)
[17]R. Chatterjee-Fischer, Boriding and Diffusion Metallizing. Surface Modification Technologies (ed. by T.S. Sudarsan), Marcel Dekker Inc., New York, USA, 1989
[18]P.A. Dearnley, T. Bell, Engineering the Surface with Boron Based Materials; *Surface Engineering*, **1**, N. 3, 203-217 (1985)
[19]H.-J. Hunger, G. Trute, Boronizing to Produce Wear-Resistant Surface Layers; *Heat Treatment of Metals,* **2**, 31-39 (1994)
[20]R. Telle, L.S. Sigl, K. Takagi, Boride-Based Hard Materials; p. 802-945 in *Handbook of Ceramic Hard Materials;* Ed. by R. Riedel, Wiley, Weinheim, 2000
[21]R.S. Petrova, N. Suwattananant; Surface Modification of Ferrous Alloys with Boron; *J. Electronic Mater.*, **34**, N. 5, 575-582 (2005)
[22]N. Suwattananant, R.S. Petrova, Oxidation Kinetics of Boronized Low Carbon Steel AISI 1018; *Oxidation of Metals*, **70**, N. 5-6, 307-315 (2008)
[23]R.S. Petrova, N. Suwattananant, V. Samardzic, The Effect of Boronizing on Metallic Alloys for Automotive Applications; *J. Mater. Eng. Perfrom.*, **17**, N. 3, 340-345 (2008)
[24]T.B. Massalski (Ed.) Handbook of Binary Alloys Phase Diagrams, 2nd edition, ASM Int. Materials Park, OH, USA, 1992
[25]H. Okamoto, B-Fe (Boron-Iron); *J. Phase Equilibrium Diffusion*, **25**, 297-298 (2004)
[26]M. Salamon, F. Fuks, H. Mehler, Interdiffusion and Al Self-Diffusion in Iron-Aluminides; *Defect and Diffusion Forum,* **237-240**, 444-449 (2005); Trans Tech Publications, Switzerland
[27]G. Palombarini, M. Carbucicchio, Growth of Boride Coatings on Iron, *J. Mater. Sci. Lett.*, **6**, N. 4, 415-416 (1987)
[28]C. Martini, G. Palombarini, M. Carbucicchio, Mechanism of Thermochemical Growth of Iron Borides on Iron; *J. Mater.* Sci., **39**, N. 3, 933-937 (2004)
[29]M. Keddam, S.M. Chentouf, A Diffusion Model for Describing the Bilayer Growth (FeB/Fe_2B) during the Iron Powder-Pack Boriding, *Appl. Surf. Sci.,* **252**, N. 2, 393-399 (2005)
[30]O. Ozdemir, M.A. Omar, M. Usta, S. Zeytin, C. Bindal, A.H. Ucisik, An Investigation on Boriding Kinetics of AISI316 Stainless Steel; *Vacuum*, **83**, N. 1, 175-179 (2008)
[31]I. Campos-Silva, M. Ortiz-Dominguez, O. Bravo-Barcenas et al, Formation and Kinetics of FeB/Fe2B Layers and Diffusion Zone at the Surface of AISI 316 Borided Steels; *Surf. Coat. Technol.*, **205**, N. 2, 403-412 (2010)
[32]V.I. Dybkov, W. Lengauer, K. Barmak, Formation of Boride Layers at the Fe-10%Cr Alloy-Boron Interface; *J. Alloys & Compounds*, **398**, 113-122 (2005)
[33]V.I. Dybkov, W. Lengauer, P. Gas, Formation of Boride Layers at the Fe-25%Cr Alloy-Boron Interface; *J. Mater. Sci.,* **41**, 4948-4960 (2006)
[34]C.M. Brakman, A.W.J. Gommers, E.J. Mittemeijer, Boronizing of Fe and Fe-C, Fe-Cr and Fe-Ni Alloys; Boride-Layer Growth Kinetics; *J. Mater. Res.,* **4**, N. 6, 1354-1370 (1989)
[35]B. Venkataraman, G. Sundararajan, The High Speed Sliding Wear Behaviour of Boronized Medium Carbon Steel; *Surf. Coat. Technol.*, **73**, N. 3, 177-184 (1995)
[36]S. Taktak, Tribological Behaviour of Borided Bearing Steels at Elevated Temperatures; *Surf. Coat. Technol.*, **201**, N. 6, 2230-2239 (2006)
[37]S. Sen, I.Ozbek, U.Sen, C. Bindal, Mechanical Behavior of Borides Formed on Borided Cold Work Tool Steel, *Surf. Coat. Technol.*, **135**, N. 2-3, 173-177 (2001)
[38]C. Martini, G. Palombarini, G. Poli, D. Prandstraller, Sliding and Abrasive Wear Behavior of Boride Coatings; *Wear*, **256**, N. 6, 608-613 (2004)
[39]M. Tabur, M. Izciler, F. Gul, I. Karacan, Abrasive Wear Behavour of Boronized AISI8620 Steel; *Wear*, **266**, N. 11-12, 1106-1112 (2009)
[40]C. Meric, S. Sahin, B. Backir, N.S. Koksal, Investigation of the Boriding Effect on the Abrasive Wear Behavior in Cast Irons; *Mater. & Design,* **27**, N. 9, 751-757 (2006)

Polymer Derived Ceramics and Composites

POLYMER-DERIVED CERAMICS FOR DEVELOPMENT OF ULTRA-HIGH TEMPERATURE COMPOSITES

C. J. Leslie[1,2], H. J. Kim[1,3], H. Chen[1,3], K. M. Walker[1], E. E. Boakye[1,3], C. Chen[1,4], C. M. Carney[1,3], M. K. Cinibulk[1], and M.-Y. Chen[1]

[1] Air Force Research Laboratory, Wright-Patterson Air Force Base, OH 45433-7817, USA
[2] NRC Research Associateship Program, Washington, DC 20001, USA
[3] UES Inc., Dayton, OH 45432, USA
[4] UDRI, Dayton, OH 45469, USA

ABSTRACT

The Ceramics Manufacturing and Processing Research Team at the Air Force Research Laboratory has employed polymer derived ceramics in the development of a number of ultra-high temperature composite systems for hypersonic flight applications. For leading edge applications, SiC fiber reinforced UHTCs have been fabricated through lamination of woven SiC fabrics and filament wound unidirectional tapes, both impregnated with HfB_2-filled SiC polymeric precursor slurries. Fabric laminates showed improved matrix density within the fiber tows compared to commercially available SiC_f/SiC preforms evaluated in previous work. Good infiltration of the laminates was achieved with HfB_2 slurry pressure infiltration followed by polymer infiltration and pyrolysis processing using a polycarbosilane SiC precursor (SMP-10) pyrolyzed at 1300°C. The resulting composite contained 30 vol% fibers and had a density of 3.97 g/cm^3. For the filament wound composites, a range of slurry formulations and heat treatments were evaluated. It was found that impregnation using a slurry of HfB_2 powder, binder and methylphenyl silicone (SR355S) polymeric SiC precursor in acetone, with a V_{HfB2}:$V_{polymers}$ ratio of 0.5, a V_{Binder}:V_{SR355S} ratio of 2, and a total solids loading of 8 vol%, a one hour autoclave cure at 425°C and 1.4 MPa pressure and pyrolysis at 1350°C resulted in relatively crack-free composites with uniform fiber distributions. For acreage thermal protection (TPS) applications, a hybrid TPS system, consisting of a carbon fiber reinforced SiC face sheet and stiffeners, silicon oxycarbide insulating foam and ceramic adhesive, all produced using polymer derived ceramics, was developed. Ceramic foams were fabricated by adding varying amounts of a foaming agent to polydimethylsiloxane, curing at room temperature, and pyrolyzing at 1200°C. Additions of up to 40 wt% SiC particles to the foams was found to decrease shrinkage during pyrolysis, but resulted in microcracking and decreased strength compared to the unfilled foams, which had strengths comparable to those seen in commercial foams. The adhesive, consisting of a SiC-loaded SMP-10, was found to have a tensile strength greater than that of the foams, ranging from 8-10 MPa.

I. INTRODUCTION

In order to realize the vision of an airbreathing hypersonic platform, a new suite of materials must be developed to meet many distinct thermal and mechanical requirements. The areas of interest are sharp leading edges, hot control surfaces, propulsion flowpaths, and acreage thermal protection (TPS). At the temperatures of interest (>1400 °C) few materials can survive and when the conditions of oxidation resistance and structural stability are applied even fewer options remain.

Additions of MeB_2 or MeC (Me = Hf, Zr and Ta) to conventional composites comprised of SiC or C fibers with SiC matrices (SiC_f/SiC or C_f/SiC) have been proposed for use in leading edges, hot control surfaces, and propulsion flow paths. The refractory borides and carbides with melting points above 3000°C are considered ultra-high temperature ceramics (UHTCs). The UHTCs are typically added as powder fillers,[1-5] produced from reactions of chemical precursors,[6] or created in situ by the reaction of a metallic phase with residual C in the matrix[7,8]. The addition of such materials at the surface and throughout the matrix helps promote higher temperature oxidation resistance[1-3,6] while maintaining the beneficial high thermal conductivity, strength, damage tolerance, and lightweight architecture of the UHTC composites (UHTCCs).

Future acreage TPS that is structurally integrated with the airframe and is therefore load-bearing, and not simply thermally protective, is desired to minimize the weight of the vehicle. Acreage TPS must be stiff and strong to maintain the shape of the airframe while at the same time being light weight. A large thermal gradient is observed from the outer structure at 1000°C to the interior of the vehicle. A current structure of interest utilizes a ceramic composite (CMC) face sheet (typically C_f/SiC or SiC_f/SiC) on top of a polymer composite (PMC) backing.[9] The two are separated in space and temperature by an insulator. The insulating material must be able to withstand the temperature gradient, as the PMC backing is limited to an operation temperature of approximately 290°C, but must also have a low density, good thermal and mechanical stability, and a thermal expansion coefficient matched to the face sheet.[10] To meet these needs SiC foams have been proposed.[10-13]

This paper explores the processing of various thermal protection materials that may be used in a hypersonic flight vehicle. The focus is on materials that benefit from the use of polymer derived ceramics and the impact of the preceramic polymers on processing windows and final properties. SiC and HfB_2 have been derived from preceramic polymers and used to process UHTCCs from both SiC fiber fabrics and tows. In addition, SiC foams were fabricated using preceramic polymer processing routes and compared to conventional foams. The structures were characterized for their mechanical strength and thermal properties.

II. EXPERIMENTAL

The two approaches for the development of SiC fiber-reinforced HfB_2-SiC CMCs investigated under this effort involve (1) the production of porous woven SiC fabric preform laminates followed by infiltration with a HfB_2 slurry and polymer infiltration and pyrolysis (PIP) processing using a SiC polymeric precursor and (2) lamination of filament wound unidirectional SiC fiber tapes impregnated with an HfB_2-loaded polymeric SiC precursor solution. These composites were characterized using scanning electron microscopy (SEM). A hybrid TPS consisting of a silicon oxycarbide insulating foam sandwiched between a carbon fabric reinforced SiC face sheet and a polymer matrix composite (PMC) face sheet was also developed. In this effort, the CMC face sheet and stiffeners, the insulating foam, and the ceramic adhesive were fabricated using the polymer-derived ceramics. The structures of the foams were characterized using X-ray computed tomography (X-ray CT) and SEM, and the mechanical properties of the ceramic foam and adhesive were evaluated using a flat-wise tension test.

II.1 Woven Fabric HfB_2-SiC Matrix Laminate Fabrication

In a previous effort, CMCs were fabricated by HfB_2 slurry infiltration of commercially available SiC_f/SiC preforms.[14] These preforms consisted of porous two and four-ply laminates of 5-harness satin weave Hi-Nicalon S fabric with a boron nitride fiber coating, partially infiltrated with SiC via chemical vapor infiltration (CVI). Although there was a significant amount of open porosity between the fiber tows that was subsequently well infiltrated by the HfB_2 matrix, the CVI SiC coating was rather thick, accounting for nearly 21% of the preform volume, and left a considerable amount of closed porosity within the tows. This resulted in a greater SiC volume than desired in the matrix and limited matrix infiltration and densification. In order to overcome these issues, partially densified SiC_f/HfB_2-SiC preform laminates were developed using woven Nicalon fabric and an HfB_2-loaded allyl-hydride-polycarbosilane (SMP-10, Starfire Systems, Inc., Malta, NY) polymeric SiC precursor. These laminates were subsequently infiltrated with an HfB_2 slurry via pressure infiltration and with SiC through multiple polymer infiltration and pyrolysis (PIP) cycles using SMP-10.

For the preparation of the preform laminates, six single plies of plain weave CG Nicalon fabric (Dow Corning, Midland, MI) were cut into 38 mm x 76 mm sections, three each with the long axis oriented in the warp and fill directions of the weave. The fabric was soaked in acetone for 2.5 hours to remove the sizing and subsequently dried in air. A slurry of SMP-10 loaded with approximately 18 vol% HfB_2 powder (-325 mesh, Materion Advanced Chemicals, Milwaukee, WI) was formulated for fabric impregnation. This formulation was selected because, based on the reported SiC yield of SMP-10, it was calculated to yield a ceramic matrix consisting of 50 vol% HfB_2 in SiC. Each fabric ply was impregnated with the slurry using standard manual fabric impregnation techniques and stacked in an alternating 0/90 (warp/fill) orientation. The resulting laminate was subsequently vacuum bagged and uniaxially pressed for 15 minutes at a pressure of 3.4 MPa in order to enhance laminate impregnation and fiber volume fraction. The laminate was heated under vacuum at 1°C/min to 250°C and held at temperature for three hours in a Lab Companion OV-11 vacuum oven (Jeio Tech Co., Ltd., Seoul, South Korea) to cure the SMP-10. The cured panel was pyrolyzed for two hours at 1300°C under flowing 99.9999% pure argon in an alumina tube furnace, resulting in a partially densified porous SiC_f/HfB_2-SiC preform similar to the SiC_f/SiC preforms used in the earlier evaluations.

The preform laminate was subsequently subjected to two pressure infiltrations using a slurry consisting of 5 vol% HfB_2 in ethanol. Each infiltration was followed by a five hour heat treatment at 1600°C in argon to partially consolidate the HfB_2 matrix. These infiltrations were followed by a series of SMP-10 PIP cycles in which the laminate was submerged in SMP-10 and vacuum infiltrated for two hours and subsequently cured and pyrolyzed using the procedures described above. After each infiltration and heat treatment, laminate mass gains were measured and approximate densities were calculated. Additionally, Archimedes density measurements were made after both the final pressure infiltration and PIP cycle. In addition to the 1300°C pyrolysis steps, a portion of the final panel was subjected to a five hour heat treatment at 1600°C. The resulting laminates were sectioned and analyzed in the scanning electron microscope (SEM, Quanta, FEI, Hillsboro, OR) to assess the tow and matrix microstructures of the composite.

II.2 Unidirectional Tape HfB_2-SiC Matrix Laminate Fabrication

Unidirectional SiC fiber tapes impregnated with an HfB_2-loaded preceramic polymer slurry were fabricated by filament winding. A range of slurries consisting of a polymeric SiC precursor (SR355S, Momentive, Columbus, OH), a binder (QPAC40, Empower Materials, New Castle, DE) and HfB_2 powder, in varying compositions shown in Table 1, was formulated for composite tape fabrication. The tapes were fabricated by first passing a Nicalon fiber tow (Dow Corning, Midland, MI) through a tube furnace at 1000°C to remove the sizing material, then through a bath of the slurry to coat the tow. The wet tow was wound into panels of the desired width on a 6-flat sided spool and allowed to dry in air. The resulting preimpregnated tape panels were then cut into pieces and stacked into 5-ply lay-ups for subsequent laminate fabrication.

Table 1. Slurry Compositions for SiC Fiber Tow Impregnation

Materials	Volume %
HfB_2	*2.5 – 4.7*
SR355S	1.6 – 6.8
QPAC40	2.0 – 5.8
Acetone	86.5 – 92.8

The laminated tapes were consolidated by pressureless sintering. For this technique, a slurry formulation consisting of an HfB_2:polymer volume ratio of 1:1 was initially used to impregnate the unidirectional tapes. The resulting plies were stacked, vacuum bagged using standard bagging techniques common to PMC processing, and heat treated under a 1.4 MPa argon pressure in an autoclave (ACS Process Systems, Chatsworth, CA). The heat treatment consisted of a 5°C/min heat up to the desired processing temperature, a one hour isothermal hold, and a subsequent 5°C/min cool down. An initial processing temperature of 95°C was chosen since this is the melting point for the SR355S precursor and was thought to be sufficient to form a green-state laminate free of delaminations and significant porosity. Based on post-processing characterization results discussed later, a modified autoclave treatment, using the same heating rates with a two-hour hold at 150°C and a one hour hold at 425°C was later adopted. With this procedure, the 1.4 MPa pressure was applied halfway through the 150°C isothermal hold. Using this modified technique, tapes fabricated using a range of slurry formulations were laminated, autoclaved and heat treated under vacuum at 1350°C in a high temperature furnace (F-14x14x14-GG-250-VM-G, MRF, Inc., Suncook, NH) to pyrolyze the SR355S into SiC. The microstructures of the resulting composites were characterized using SEM.

II.3 CMC/PMC Hybrid Fabrication

In this hybrid TPS, preceramic polymers are used in the fabrication of three types of structural elements: the CMC face sheet and stiffeners, the ceramic foam, and the ceramic adhesive constructing the structure. The face sheet and stiffeners were fabricated using the slurry impregnation and lamination technique discussed earlier. In a typical process, plies of pyrolytic carbon coated plain weave carbon IM7 fabric were impregnated with a slurry of SMP-10 loaded with 30 vol% SiC powder (Materion Advanced Chemicals, Milwaukee, WI). A total of 6 plies were then laminated in an alternation 0/90 (warp/fill) orientation, cured at 250°C under vacuum and pyrolyzed to 1200°C in argon.

The ceramic foam was fabricated via pyrolysis of a silicone foam. Preparation of the silicone foam started with mechanical mixing of vinyl terminated polydimethylsiloxane (Aldrich, Milwaukee, WI), foaming agent, platinum catalyst, and crosslinking agent. Mechanical stirring served to both homogenize the mixture and initiate air bubbles to serve as nuclei for foam cell growth. Additionally, up to 40 vol% SiC powder was added to some of the foam formulations as a filler to minimize shrinkage during subsequent pyrolysis. Foaming and curing initiated instantly upon mixing at room temperature. Typically, the liquid foam fully solidified within 10 minutes due to catalyzed crosslinking. Prior to solidification, the foam can be cast into a mold for fabricating shaped components. The porosity, cell size, and density of the foam can be controlled by the contents of blowing agent and SiC filler particles. The silicone foams were pyrolyzed at a peak temperature of 1200°C in argon. The densities of the cured silicone foams and their derived ceramic foams were measured and their structures were analyzed using 3D image analysis on the X-Ray CT (computed tomography) scan data obtained from an X-TEK HMX 160 system (Nikon Metrology, Brighton, MI). The porosity, cell size distribution and cell number density were measured to study the effect of blowing agent and filler content. Finally, foam specimens (12.7 mm x 12.7 mm x 6.4 mm) were mounted and tested in custom flat-wise tension configuration to determine the effects of foam structure, density and filler addition on the tensile strength of the foams.

Assembly of the TPS structure was achieved by the application of a preceramic polymer based adhesive. The adhesive was prepared by dispersing SiC particles in SMP-10 using a planetary vacuum mixer (ARV-310, Thinky USA, Laguna Hills, CA). SiC particles of two different particle sizes, 5 μm and 1 μm, were used at a ratio of 7:3 to maximize the packing density. Solid particle loadings between 20 vol% and 60 vol% were evaluated. The adhesive was applied to the CMC and the foam surface and the bonded structure was pyrolyzed to 1200 °C in flowing argon. The bond strength of the bonded specimens (12.7 mm x 12.7 mm bonded area) was evaluated using a flat-wise tension configuration.

III. RESULTS AND DISCUSSION

III.1 Woven SiC Fabric Reinforced UHTCs

In order to avoid the closed porosity seen in the previously used SiC_f/SiC preforms caused by the use of an initial CVI densification step, SiC fabric laminates were prepared using slurries of SMP-10 loaded with approximately 18 vol% HfB_2 powder. The SiC laminates consisted of a six ply layup of plain weave Ceramic Grade Nicalon fabric with a matrix of HfB_2-SiC within the fiber tows. Similar to the SiC_f/SiC preforms, the laminate had an open structure, with significant open porosity in the inter-tow regions. The fiber volume fraction and laminate density were calculated at 30 vol% and 1.68 g/cm^3, respectively. These values are quite comparable to the values of 37 vol% and 1.9 g/cm3 reported for the SiC_f/SiC preforms. Given that much of the density of the SiC_f/SiC preforms was attributable to the thick CVI SiC coating, the lower density of the preform laminate suggested a lighter ceramic matrix around the tows and greater porosity.

SEM backscatter images of the fiber tow in the commercial preform, fabricated using CVI, and the preform laminate fabricated through SMP-10 slurry impregnation are shown in Fig. 1. From Fig. 1b, it is evident that a significant percentage of the closed porosity within the tow was eliminated by the polymer impregnation process. Although there was observed closed porosity within many of the tows, these pores were typically smaller and less prevalent than those seen in the CVI preforms. It is also apparent that the matrix composition within the tows was improved by the polymer impregnation technique, with the clear presence of HfB_2 dispersed in the SiC matrix within the tows. There were, however, regions within the tows that were quite SiC-rich, as seen in the upper portion of Fig. 1b. This was likely due to a filtering effect of the fibers by which further ingress of the HfB_2 particles was blocked while the liquid SiC precursor was allowed to flow through. Improvements in the impregnation process and incorporation of smaller HfB_2 particles should alleviate this matrix constituent distribution issue.

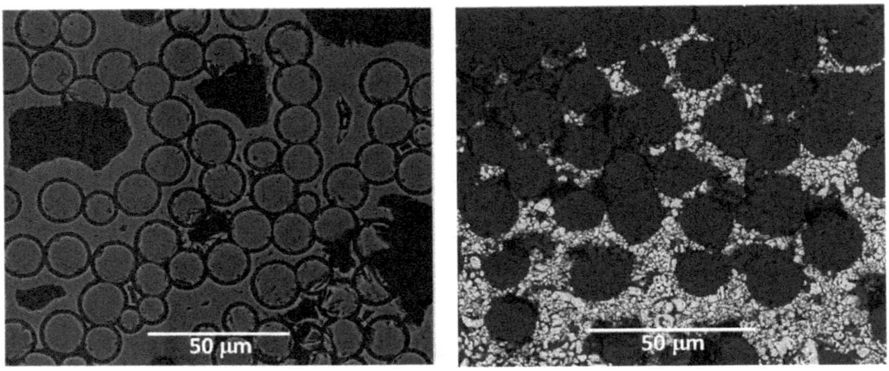

(a) (b)

Fig. 1. SEM backscatter micrographs of a fiber tow in the (a) commercial SiC_f/SiC preform with excessive SiC matrix and (b) fabricated SiC_f/HfB_2-SiC preforms. The bright phase is HfB_2 while the grey phase is SiC.

After fabrication, the preform laminates were subjected to two HfB_2 slurry pressure infiltrations and five SMP-10 PIP cycles. The microstructure of the final composite is shown in Fig. 2. It is apparent, from Fig. 2a, that good infiltration was achieved during slurry infiltration. The micrograph in Fig. 2b indicates that appreciable infiltration of the matrix porosity was achieved through PIP processing, with the presence of a continuous amorphous SiC matrix with HfB_2 dispersed throughout after heat treatment at 1300°C. However, the SiC crystallized during the 1600°C heat treatment, resulting in an increase in matrix cracking and porosity (Fig. 2c).

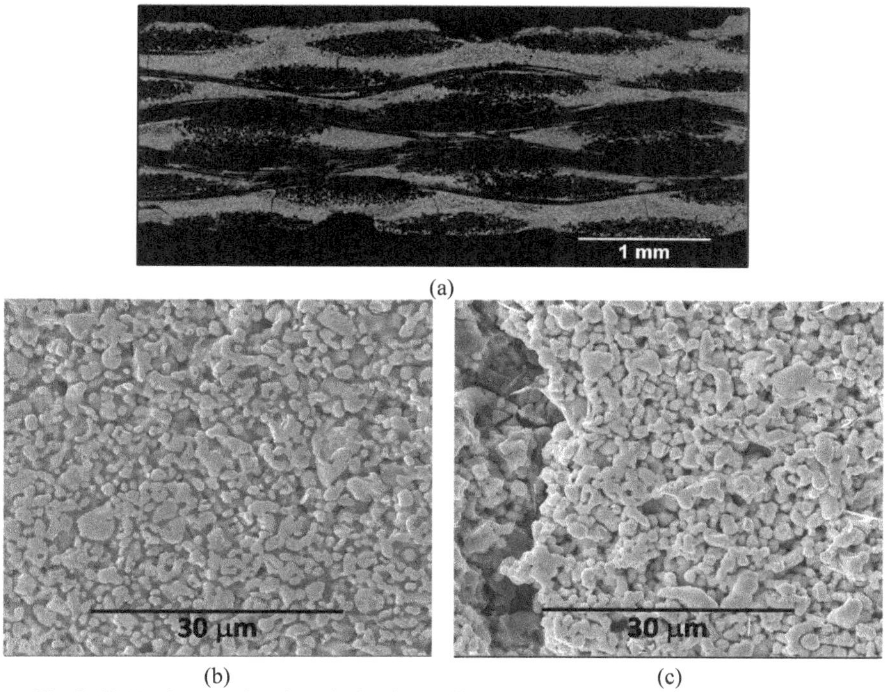

Fig. 2. SEM micrographs of (a) the laminate after pressure infiltration and PIP processing (backscatter), and the matrix after final heat treatment at (b) 1300°C and (c) 1600°C.

After each infiltration step, the mass of the laminate was measured and the density was estimated using volumes calculated from dimensional measurements. The progression of densification with each processing step is shown in Fig. 3. As expected, the most significant density increase occurred during the slurry infiltrations when the HfB_2 component of the matrix was added. During these steps, the composite density increased by over 80%. Repeated PIP cycles resulted in additional matrix densification which gradually decreased to a negligible level after five cycles, with a density increase of only 1% after the fifth cycle. The final density, measured using the Archimedes method, was 3.97 g/cm^3. This represents an overall density increase of over 136%. However, it is considerably lower than the theoretical density of 7.09 g/cm^3 determined based on the fiber volume fraction and desired matrix composition. This disparity was the result of both a deviation in the matrix composition from the desired HfB_2 content and the presence of closed porosity within the matrix.

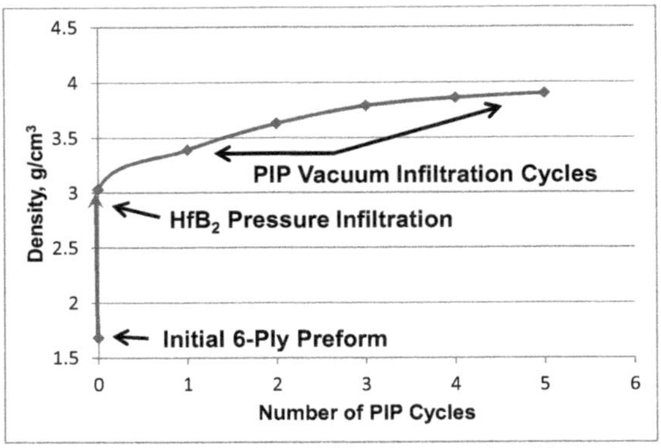

Fig. 3. Densification of SiC_f/HfB_2-SiC laminate during fabrication process.

III.2 Filament Wound SiC Fiber Reinforced UHTCs

Initially, a slurry formulation consisting of an HfB_2-to-polymer volume ratio (V_{HfB2}:$V_{polymers}$) of one, a QPAC40-to-SR355S volume ratio (V_{QPAC40}:V_{SR355S}) of 0.3, and a total solids loading (HfB_2 plus solid polymers) of 16 vol% in acetone was used to fabricate the filament wound composite. These tapes were laminated and autoclaved using both the initial and modified protocols discussed above prior to 1350°C pyrolysis. Micrographs showing the resulting microstructures are shown in Fig. 4. As shown in the micrograph in Fig. 4a, the microstructure of the composite produced using the initial autoclave protocol contained large pores and showed evidence of uneven laminate compression, despite the application of the 1.4 MPa pressure. In order for the particles and fibers to more effectively flow and rearrange under the applied pressure, the higher temperature modified protocol was adopted. The increased temperature of this protocol was expected to improve composite microstructure by reducing the viscosity of the molten polymers, thus allowing them to behave as lubricants for fiber and particle flow. As shown in Fig. 4b, although laminate compression appeared to show slight improvement with the modified protocol, the composite still contained a significant amount of large pores. In both cases, sizable regions of unreinforced matrix between plies were also observed.

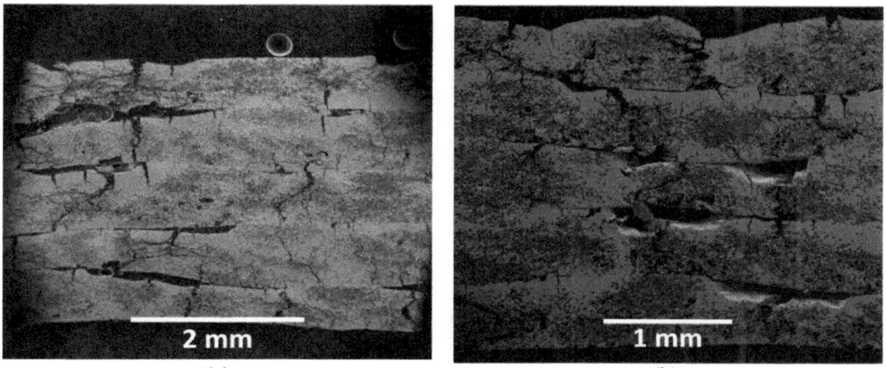

(a) (b)

Fig. 4. Microstructure of composites processed using (a) the initial autoclave protocol and (b) the modified protocol. Volume ratios of the slurry components were V_{HfB2}:$V_{polymers}$ = 1 and V_{QPAC40}:V_{SR355S} = 0.3, with a total solids loading of 16 vol%.

Since both the temperature and pressure of the modified protocol were at the operation limits of the autoclave, the focus for further improving the laminate microstructures was shifted to the slurry composition. Initially, the V_{HfB2}:$V_{polymers}$ ratio was 1. In order to further improve laminate quality, a new laminate was fabricated using a formulation in which the V_{HfB2}:$V_{polymers}$ ratio was reduced to 0.5 by the addition of polymers. The microstructure of the resulting composite is shown in Fig. 5. As these micrographs reveal, the large pores were successfully removed and laminate compression was dramatically improved with the new formulation. During the 1350°C pyrolysis, however, a significant number of cracks were generated. Generally, cracks are generated by shrinkage mismatch among the components in a composite. Sintering shrinkage of HfB_2 and Nicalon fibers is negligible because 1350°C is insufficient for sintering of these materials. The only component in the system that exhibits a significant volume change is the polymer. The QPAC40 should be removed during autoclave processing, thus the only polymer present during pyrolysis is SR355S. The degree of shrinkage in the unreinforced matrix is expected to differ from that in the dense fiber region during heat treatment due to the shrinkage of the SR355S between the HfB_2 particles. This shrinkage difference can be reduced by minimizing the amount of residual SR355S present in the green-state composite.

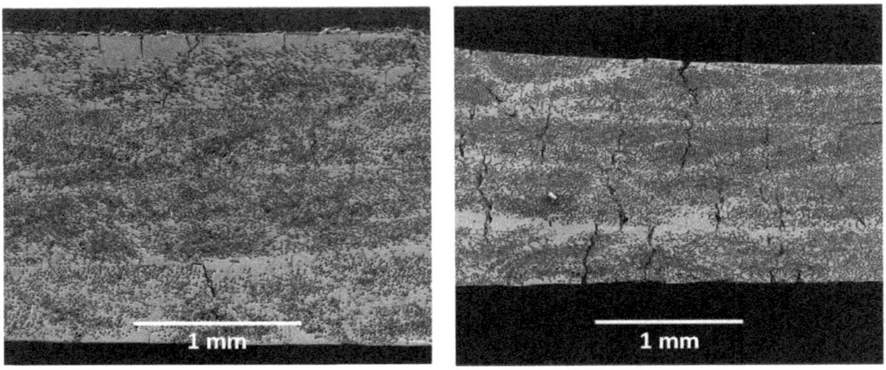

(a) (b)

Fig. 5. Microstructure of composites processed using the modified slurry composition (a) after the autoclave run and (b) after 1350°C pyrolysis. Volume ratios of the slurry components were V_{HfB2}:$V_{polymers}$ = 0.5 and V_{QPAC40}:V_{SR355S} = 0.3, with a total solids loading of 16 vol%.

To minimize the size and number of the cracks, further modification of the slurry composition was made. Keeping the total amount of polymers constant, the amount of QPAC40 was significantly increased to a V_{QPAC40}:V_{SR355S} ratio of 2. The resulting microstructure, shown in Fig. 6, possessed significantly smaller cracks than the previous laminate after pyrolysis. However, the regions of unreinforced matrix between plies remained.

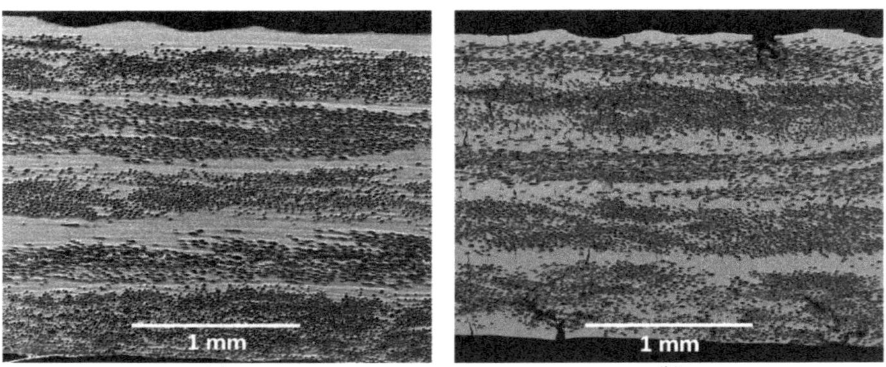

(a) (b)

Fig. 6. Microstructure of composites processed using the modified slurry composition (a) after the autoclave run and (b) after 1350°C pyrolysis. Volume ratios of the slurry components were V_{HfB2}:$V_{polymers}$ = 0.5 and V_{QPAC40}:V_{SR355S} = 2, with a total solids loading of 16 vol%.

In order to eliminate these regions, the amount of acetone in this formulation was varied to control the viscosities of the slurries which, in turn, controlled the amount of dip-coated slurry on the tows during winding. Fig. 7 shows the microstructures of the composites prepared with slurries containing various amounts of acetone. Near complete removal of these unreinforced regions was achieved with a total solids loading of 8 vol%. From these results, a slurry formulation with a V_{HfB2}:$V_{polymers}$ ratio of 0.5, a V_{QPAC40}:V_{SR355S} ratio of 2, and a total solids loading of 8 vol% and the modified autoclave protocol were selected as the optimum composite processing parameters to be used for laminate fabrication going forward.

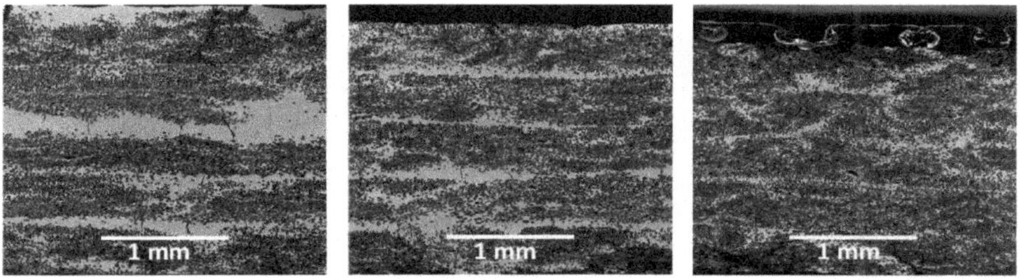

(a) (b) (c)

Fig. 7. SEM micrographs of the composites after autoclave and pyrolysis prepared with slurries with total solids loadings of (a) 14 vol%, (b) 11 vol%, and (c) 8 vol%.

III.3 CMC and Ceramic Components for Hybrid TPS

In the development of the hybrid TPS, preceramic polymers were used in the fabrication of CMCs, the insulating ceramic foam, and the high temperature adhesive. Since CMC development for this system was nearly identical to the process used for the woven fiber reinforced UHTC laminate development, only the insulating foam and adhesive development will be discussed here.

The ceramic foam was fabricated via pyrolysis of a silicone foam. Preparation of the silicone foam started with mechanical mixing of vinyl terminated polydimethylsiloxane, foaming agent, platinum catalyst and crosslinking agent. Variations in the SiC filler content and foaming agents were explored to determine the best processing methods.

It was observed that the unfilled silicone foams tended to crack after pyrolysis. Additions of 15 wt% SiC particles to the initial foaming mixture, however, yielded crack-free ceramic foams. Further additions, up to 40 wt% SiC particles, were found to further reduce the volume shrinkage during pyrolysis without significantly affecting the foaming process. Since this formulation produced crack-free foams with minimal shrinkage, a 40 wt% particle loading was used for subsequent development work.

The effect of the foaming agent additions on the structure of the silicone foams with 40 wt% SiC filler and derived silicon oxycarbide foams, reconstructed from X-Ray CT scan data, is shown in Fig. 8. When no foaming agent was used, closed pores were initiated during the mechanical stirring of the silicone precursor. As foaming agent content increased to 2%, the porosity generally increased, corresponding to a decrease in density. Higher concentrations of foaming agents resulted in increasingly smaller cell population density accompanied by insignificant change in foam density, suggesting that cell merging and collapsing took place due to excessive cell growth. The cells are more interconnected when the foam becomes more porous, which can affect the insulation capability of the material. However, unlike commercial open-cell SiC foams synthesized from polymer foam templates, these silicone derived ceramic foams prepared using the direct foaming method maintain a percentage of closed porosity.

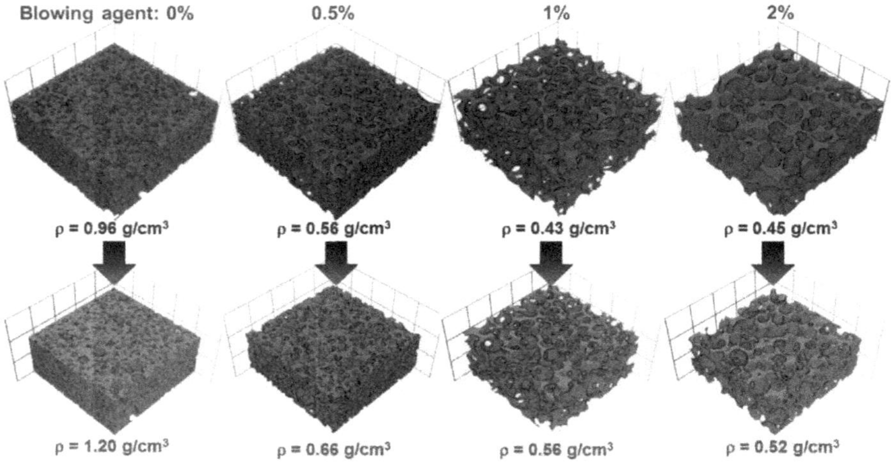

Fig. 8. X-ray CT results showing the effect of foaming agent additions on the structure and density of the silicone foam with 40 wt% SiC before pyrolysis (top) and the pyrolyzed silicon oxycarbide foam (bottom). Scale: each grid square is 1 mm x 1 mm.

From the images in Fig. 8, it is evident that the foams underwent approximately 20% linear shrinkage during pyrolysis. It has been mentioned earlier that with 40 wt% SiC additions, the shrinkage was reduced compared to silicone foams with lower SiC filler content. Although the presence of these fillers decreased shrinkage, Fig. 9 shows that it also resulted in an undesired increase in density and decrease in foam strength. The struts of ceramic foam derived from unfilled silicone are solid without visible pores or cracks (Fig. 9a). For the filled silicone foams, shrinkage of the matrix due to silicone-to-ceramic transformation led to microcracking around the rigid particles (Fig. 9b). As Fig. 9c shows, flatwise tensile testing revealed that the filler additions resulted in a decrease in the specific strength of the ceramic foams, while the unfilled foams showed strength comparable to that of the commercial Ultramet SiC foams of similar density, which were synthesized at a higher temperature. TGA analysis of silicone derived ceramic foams showed good thermal stability, with negligible weight change when heated to between 1100°C and 1400°C in argon.

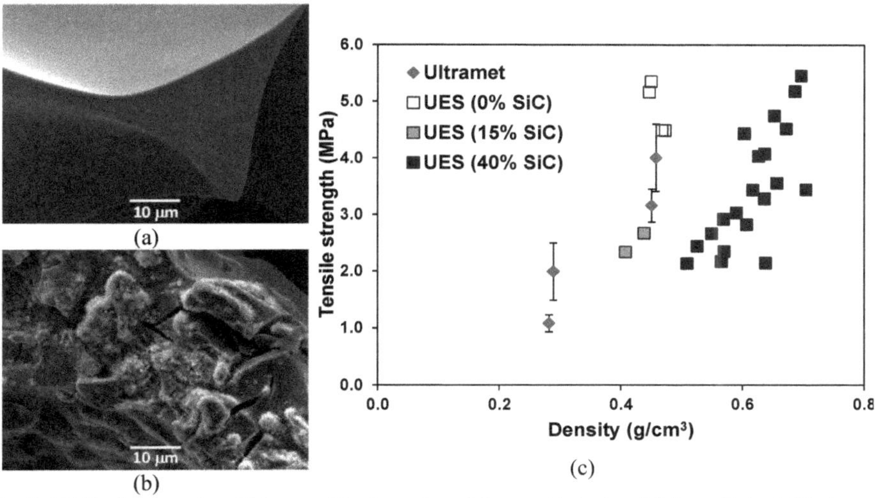

Fig. 9. SEM micrographs of (a) an unfilled pyrolyzed foam cell strut and (b) a cell strut filled with 31% SiC and (c) the effects of filler additions on tensile strength of the foams.

Finally, the SMP-10-based adhesive was evaluated. The performance of this adhesive showed a strong dependence on the SiC particle loading. For particle loadings below 50 vol%, the bonded components debonded immediately after pyrolysis due to the excessive shrinkage and cracking of the bond layer. The optimum bond strength was achieved with a SiC solid loading between 55 vol% and 60 vol%. The bond strength, tested between monolithic SiC plates using the flatwise tension method, was measured at 8 – 10 MPa. When the adhesive was used to bond the ceramic foams to CMC, failure always occurred in the foam since the majority of foams tested were weaker than the adhesive.

IV. CONCLUSION

There are numerous ongoing efforts at AFRL to develop a diverse range of ceramics and CMCs from polymer-derived ceramics to address the current and projected materials performance requirements of the Air Force. Development of continuous SiC fiber reinforced UHTC materials for hypersonic leading edge applications through the lamination of woven Nicalon fabric and preimpregnated unidirectional tapes has made significant progress in its initial phases, with mechanical and thermal properties characterization forthcoming. A hybrid TPS system, incorporating a C_f/SiC face sheet adhered to a silicon oxycarbide insulating foam with a ceramic adhesive, all produced using polymer-derived ceramics, is being developed. The ceramic components have been evaluated and modeled airframe structures incorporating this system are being fabricated for further evaluation.

ACKNOWLEDGEMENTS

The authors wish to acknowledge K. A. Keller (UES, Inc.) and R. Corns (UES, Inc.) for their technical support. This research was performed while C.J.L. held a National Research

Council Research Associateship Award at the Air Force Research Laboratory, Wright-Patterson Air Force Base, OH. H.J.K, H.C., E.E.B. and C.M.C. were supported under Air Force Contract No. FA8650-10-D-5226, and C.C. was supported under Air Force Contract No. FA8650-10-D-5011.

[1] A. Ortona, P. Fino, C. D'Angelo, S. Biamino, G. D'Amico, D. Gaia and S. Gianella, Si-SiC-ZrB_2 ceramics by silicon reactive infiltration, *Ceram. Int.*, **38**, 3243-50 (2012).
[2] A. Paul, S. Venugopal, J. G. P. Binner, B. Vaidhyanathan, A. C. J. Heaton and P.M. Brown, UHTC-carbon fibre composites: Preparation, oxyacetylene torch testing and characterization, *J. Eur. Ceram. Soc.*, **33**, 423-32 (2013).
[3] S. Tang, J. Deng, S. Wang, W. Liu, and K. Yang, Ablation behaviors of ultra-high temperature ceramic composites, *Mater. Sci. Eng. A*, **465**, 1–7 (2007).
[4] D. Zhao, C. Zhang, H. Hu and Y. Zhang, Preparation and characterization of three-dimensional carbon fiber reinforced zirconium carbide composite by precursor infiltration and pyrolysis process, *Ceram. Int.*, **37**, 2089–93 (2011).
[5] S. R. Levine, E. J. Opila, M. C. Halbig, J. D. Kiser, M. Singh and J. A. Salem, Evaluation of ultra-high temperature ceramics for aeropropulsion use, *J. Eur. Ceram. Soc.*, **22**, 2757–67 (2002).
[6] D. Zhao, C. Zhang, H. Hu and Y. Zhang, Ablation behavior and mechanism of 3D C/ZrC composite in oxyacetylene torch environment, *Compos. Sci. .Technol.*, **71**, 1392-96 (2011).
[7] Y. Wang, X. Zhu, L. Zhang, L. Cheng, C/C-SiC-ZrC composites fabricated by reactive melt infiltration with $Si_{0.87}Zr_{0.13}$ alloy, *Ceram. Int.*, **38**, 4337-43 (2012).
[8] L. Zou, N. Wali, J-M. Yang and N. Bansal, Microstructural development of a C_f/ZrC composite manufactured by reactive melt infiltration, *J. Eur. Ceram. Soc.*, **30**, 1527-35 (2010).
[9] C. A. Steeves and A. G. Evans, Optimization of Thermal Protection Systems Utilizing Sandwich Structures with Low Coefficient of Thermal Expansion Lattice Hot Faces, *J. Am. Ceram. Soc.*, **94**, S55–S61 (2011).
[10] F. I. Hurwitz, W. S. Steffier, J. R. Koenig and J. D. Kiser, Improved fabrication of ceramic matrix composite/foam core integrated structures, *NASA Tech Briefs* 2009:36.
[11] A. Ortona, S. Pusterla and S. Gianella, An integrated assembly method of sandwich structured ceramic matrix composites *J. Eur. Ceram. Soc.*, **31**, 1821–26 (2011).
[12] T. Pichon, R. Barreteau, P. Soyris, A. Foucault, J. M. Parenteau, Y. Prel, et al., CMC thermal protection system for future reusable launch vehicles: generic shingle technological maturation and tests, *Acta Astronaut*, **65**, 165–76 (2009).
[13] P. Colombo, Conventional and novel processing methods for cellular ceramics, *Philos. Trans. R. Soc. A*, **364**, 109–24 (2006).
[14] C. J. Leslie, E. E. Boakye, K. A. Keller and M. K. Cinibulk, Development of Continuous SiC Fiber Reinforced HfB_2-SiC Composites for Aerospace Applications, *Materials Science and Technology 2012 Conference and Exhibition*, Pittsburgh, PA, October 7-11, 2012.

SILICONBORONOXYCARBIDE (SIBOC) FOAM FROM METHYL BOROSILOXANE

Sreejith Krishnan, Tobias Fey*, and Peter Greil
University of Erlangen-Nuremberg,
Department of Materials Science (Glass and Ceramics), Erlangen, Germany
*E-mail: tobias.fey@fau.de

ABSTRACT

Siliconboronoxycarbide (SiBOC) foams were prepared from borosiloxane precursors by sacrificial pu foam template method. Methyl borosiloxane oligomer was synthesized from boric acid and methyltriethoxysilane. This SiBOC precursor (methylborosiloxane) was pyrolyzed at 1000°C in argon with a ceramic yield of amorphous SiBOC of ~78%. Compressive strength of the foam sample achieved values of 410kPa at a total porosity of 85% and a mean pore cell diameter of approx. 700 µm. Annealing at 1400°C in argon resulted in β-SiC crystallization and a SiC/SiBOC nano composite foam was formed.

INTRODUCTION

Ceramic foam materials have gained much momentum over the past few years. Low density, high surface area, high permeability and low specific heat have enabled these materials to find application as catalyst supports, filters for molten metals and hot gases, thermal protection materials and heat exchangers[1]. The production of ceramics from the pyrolysis of filler free polymeric precursors has already claimed a pivotal position as it provides many advantages over the conventional methods. The advantages include low temperature processability, high purity of the ceramics, the ability to control the composition of the ceramics at the molecular level, easiness in the preparation of shaped objects and the ability to generate products in a variety of forms such as coatings, films, joints, monolithic bodies, foams and fibers[2]. The preparation of ceramics from polymers involves the following steps:

a) synthesis of preceramic polymers from monomers

b) shaping and curing of green body at temperatures <250°C

c) pyrolysis in the temperature range 500 to 1600°C under inert or reactive gas.

During pyrolysis decomposition of cross-linked polymer takes place in the temperature range 400 to 800°C, followed by conversion to amorphous ceramic and crystallization of the ceramic occurs in the temperature range 1300 and 1600°C. The ceramics obtained by polymer pyrolysis is amorphous in nature owing to its unique chemical composition and the low processing temperature. For crystallizing this amorphous ceramics, high temperature annealing is required. One of the advantages of the polymer-derived ceramic materials is their ability to remain amorphous at high temperatures thereby retaining their mechanical properties. Detailed micro-structural / compositional analysis of the ceramics shows that the transformation of amorphous ceramics to crystalline material is not via purely crystalline intermediates. The materials are in-homogeneous in phase distribution consisting segregates of metastable intermediates[3]. With increase in temperature, size of these metastable phases increases and leads to the formation of a crystalline material. The type of metastable phase formed during pyrolysis depends strongly on the chemical composition of the precursor. This is the major advantage of precursor route to ceramics that the composition and properties of the ceramics can be defined by the structure and composition of the polymer.

Among the polymer derived ceramic materials, siliconoxycarbides (SiOCs) in particular, are the most investigated material due to their good mechanical strength, creep resistance, high temperature stability and resistance to oxidation and corrosion. SiOC foams are therefore widely explored for high temperature applications. They are produced from siloxane resins by pyrolysis at 1000°C under inert atmosphere. SiOC structure is a random network of silicon-oxygen tetrahedra, with some silicon bonded to one or two carbons substituted for oxygen; these carbons are in turn tetrahedrally bonded to other silicon atoms. There are very small regions of carbon-carbon bonds only, which are not bonded in the network, known as the free carbon phase[4,5].

Siliconboronoxycarbides (SiBOCs) are a new class of mixed oxycarbides having interesting phase stability characteristics and thermooxidative stabilities[6,7]. SiBOC glasses are thermally stable in the range 1000–1500°C. In the same temperature range, the crystallization of nano-sized β-SiC takes place. The tendency towards β-SiC crystallization seems strongly dependent on the boron load, with a decrease of the crystallization temperature and an increase in β-SiC crystal size the B/Si ratio. During the β-SiC crystallization, the mixed silicon and boron oxycarbide units are consumed and that, at the end of the process, boron is mainly present in trigonal borosilicate ($B(OSi)_3$) sites. Such an effect of boron on the phase stability and β-SiC crystallization can be of great use in designing ceramic compositions with variable properties. The SiBOC ceramic when exposed to oxidizing environment at high temperature, forms a protective borosilicate glassy layer on the surface, which prevents further oxidation of the ceramic[8]. Borosiloxanes are the most common precursors for SiBOC ceramic compositions. Polyborosiloxanes have been an exploration target because of their thermodynamically stable B-O-Si backbone linkage, despite their high susceptibility to hydrolysis[9]. Among the inorganic polymers, boron and silicon containing polymers generally have superior thermo-chemical properties[8,10]. The high thermal stability of these polymers is attributed to the high dissociation energies of Si-O and B-O bonds and the higher extend of crosslinking.

As ceramic precursors, till now, polyborosiloxanes were investigated mainly for use as binders for making SiC and Si_3N_4 components[11], SiBOC glass fibers[12] and as matrix precursors for the preparation of ceramic matrix composites [13,14]. Recently there are reports on strong broad band tunable visible emission of SiBOC ceramic films[15].

In the present study SiBOC based foams were prepared from borosiloxane precursor. The SiBOC foam obtained is characterized for its composition, morphology and compressive strength. The SiBOC foam was converted into SiC/SiBOC nano composite foam after heat treatment at 1400°C.

EXPERIMENTAL

Materials:
Open cell reticulate polyurethane (PU) foam with a cell density of 45ppi (Panta-Cell, Koepp-Schaum, Oestrich-Winkel, Germany) was used as the sacrificial template for the preparation of SiBOC foam. For the synthesis of borosiloxane precursor, Boric acid (VWR International, Dublin, Ireland, 99.8%) and methyltriethoxysilane [MTEOS] (ABCR Gmbh & Co. KG, Karlsruhe, Germany, 98%) were used as received.

Preparation of SiBOC foam from borosiloxane:

The methylborosiloxane precursor was synthesized from boric acid and methyltriethoxysilane following the non-aqueous synthesis method[16]. The processing of SiBOC foam from PU template was carried out following the common Schwartzwalder process[16]. The polyurethane template impregnated by dip-coating (three times) with the precursor. The borosiloxane coated PU foam was kept in an air oven at 125°C for 5 hrs for the evapouration of ethanol and subsequent crosslinking of the precursor. Conversion to SiBOC was carried out by pyrolysis at 1000°C under argon flow in a tubular alumina furnace with a heating and cooling rate of 3°K/min with a dwell time of 3 hrs. Similar procedure was followed for the further heat treatment of the SiBOC foam at 1400°C.

Characterization:

Thermal degradation of Methyl borosiloxane oligomer and polyurethane template was followed by thermogravimetric analysis under argon atmosphere (STA 409, Netzsch Geraetebau, Selb, Germany). The ceramic conversion was followed by XRD (Siemens Krystalloflex D500, Mannheim, Germany) and FT-IR (Nicolet Impact 420 with OMNIC ESP 5.1) analyses. B content of the polymer and ceramic samples were analyzed by ICP-AES (Spectro Genesis, Spectro Analytical Instruments Gmbh, Kleve, Germany). Si content of the polymer and ceramic samples were determined by gravimetry.

The morphology of the SiBOC foam samples were analyzed by SEM (ESEM, Quanta 200 FEG, FEI, Eindhoven, Netherlands). Compressive strength of the SiBOC foam was measured using Instron UTM (Instron 5565, Instron GmbH, Pfungstadt, Germany). The geometrical density of the SiBOC foam was measured from the weight and volume. Strut densities were determined by He gas pycnometry (Accupyc 1330, Micromeretics, Moenchengladbach, Germany) of powdered SiBOC foam. The specific surface area was measured from N_2 sorption isotherms according to the Brunauer–Emmett–Teller (BET) method (ASAP 2000, micromeretics, Moenchengladbach, Germany). To reveal the microstructure, the samples were examined with a high resolution micro computed tomography (μ-CT) analysis using a Skyscan 1172, (Skyscan B.V., Kontich, Belgium) with a 11 MP detector. The tungsten tube was operated at a voltage of 80 kV and a current of 100 μA using no additional source filtering. The scan was performed with a rotation step of 0.4° over 180° with a resolution of 6.9 to 8 μm / voxel depending on sample size.

Measured raw data sinograms were reconstructed with the tomographic reconstruction software (NRecon Client and Server 1.6.8 with GPU support, Skyscan, Kontich, Belgium), which calculates the two-dimensional (2D) cross-sections after adjusting grey value levels. The 3D visual images and STL-models were generated using imaging software (Amira 5.4.1, Visage Imaging, Berlin, Germany) after labeling with a global threshold and using a 26-side growing algorithm on all layers.

RESULTS AND DISCUSSION

Synthesis and Characterization of methylborosiloxane:

The methylborosiloxane precursor has the basic structural unit containing B-O-Si bonding with the methyl group attached to Si, as shown in Fig. 1.

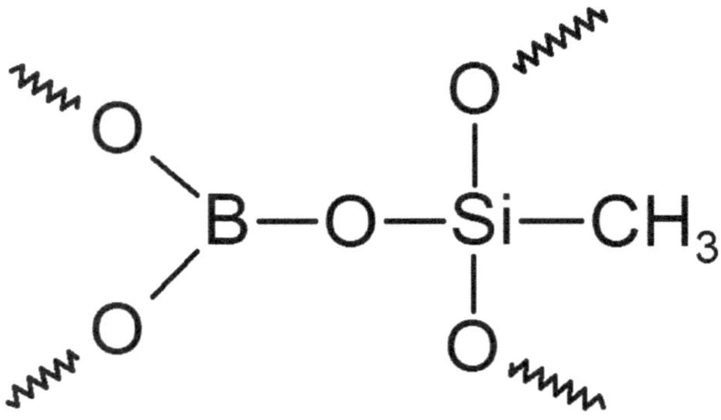

Fig. 1. Basic structural unit in methylborosiloxane.

The precursor was obtained as a solution in ethanol, which is produced as the reaction byproduct. The formation of B-O-Si bond in the precursor is proved by the B-O-Si stretching and bending vibrations recorded in the IR spectra of the precursor (Fig. 2).

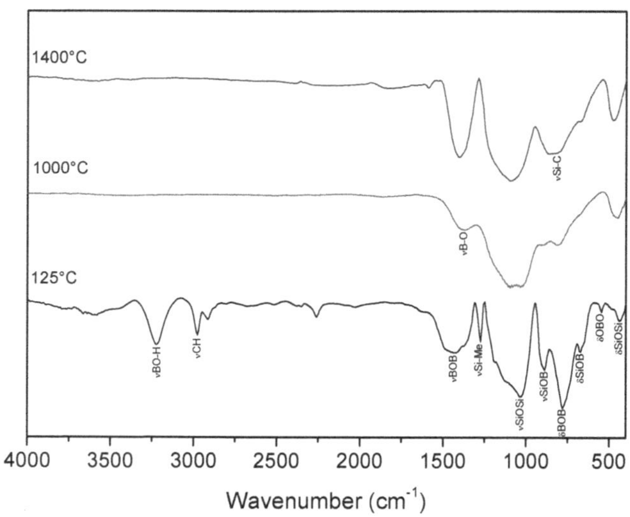

Fig. 2: IR spectra of methylborosiloxane at different stages of ceramic conversion.

The complete peak assignment is presented in Table 1.

Table 1: Peak assignment.

Wavelength (cm^{-1})	**Assigned functional group**
439	δO-Si-O
547	δO-B-O
779	δB-O-B
887	νSi-O-B
1033, 1121	νSi-O-Si
1133	νSi-O-C (of unreacted Si-O-Et).
1187	δB-OH
1272	$\nu Si\text{-}CH_3$
2260	B-OH combination band
2977, 2915	νC-H
3224	νO-H of B-OH

The presence of a peak at 887 cm^{-1} corresponding to the stretching vibration of B-O-Si supports the formation of B-O-Si linkages. The sharp absorption observed at 1272 cm^{-1} corresponds to ν $Si\text{-}CH_3$. A broad peak at 1033 cm^{-1}, which has merged with the peak at 1187 cm^{-1}, is attributed to the stretching vibration of Si-O-Si linkages. This observation suggests that the precursor contains Si-O-Si linkages in addition to B-O-Si linkages.The peak at 1133 cm^{-1} corresponding to stretching vibration of Si-O-C (of unreacted Si-O-Et). In addition to these, the absorptions seen at 3224 cm^{-1} corresponding to O-H stretching of B-OH. The above observations suggest that unreacted B-OH and Si-OEt groups are also present. The chemical reactions which are responsible for the formation of Si-O-B and SiOSi bonding are shown in Table 2.

Table 2: Possible reactions

=B-OH + ≡Si-OR	→	=B-OR + ≡Si-OH	(i)
=B-OH + ≡Si-OH	→	=B-O-Si≡ + H_2O	(ii)
=B-OR + ≡Si-OH	→	=B-O-Si≡ + ROH	(iii)
=B-OH + ≡Si-OR	→	=B-O-Si≡ + ROH	(iv)
≡Si-OH + ≡Si-OH	→	≡Si-O-Si≡ + H_2O	(v)
≡Si-OH + ≡Si-OR	→	≡Si-O-Si≡ + ROH	(vi)

The TG curves of borosiloxane oligomer (Fig. 3) shows three different degradation regions.

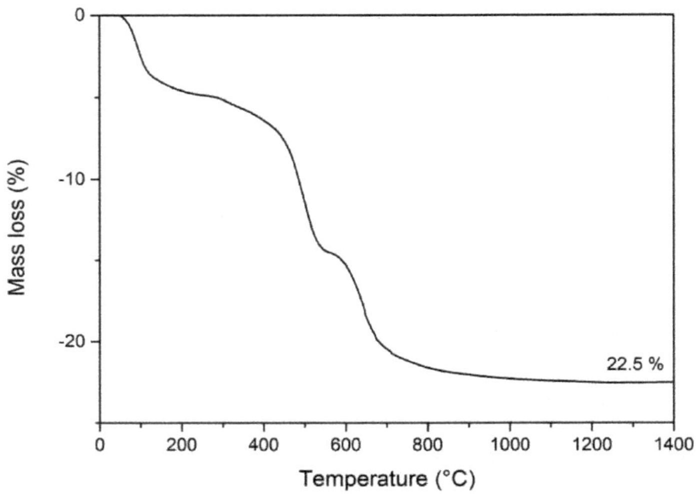

Fig. 3: TG curve of methylborosiloxane.

The oligomer has given a ceramic residue of 78 % at 1000°C and at 1400°C under inert atmosphere. The first region at RT-250°C, is the lowest one which accounts for 5 Mass% of the total weight loss, which can be attributed to the loss of low molecular weight fractions. The second region at 250-525°C corresponds to 8.5 Mass% weight loss and is attributed to the pyrolysis loss of organic groups and the third degradation region at 525-1000°C contributes 8.8 Mass% and is attributed to the ceramization process. From 1000-1400°C the borosiloxane showed almost no weight loss (<0.2 Mass%) proving the high thermal stability of the SiBOC composition at this temperature regime.

The ceramic conversion of the methylborosiloxane at 1000°C was characterized by XRD and IR analyses, see spectra Figs. 2 and 4 respectively.

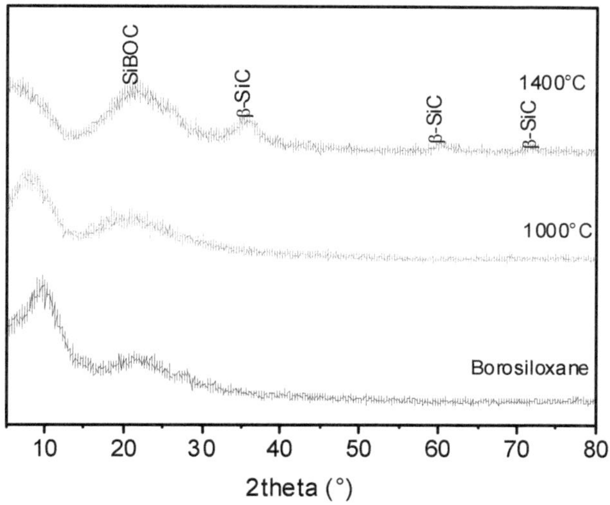

Fig. 4: XRD spectra of methylborosiloxane at different stages of ceramic conversion.

The XRD pattern shows a diffraction halo centered around 2θ = 22°, corresponding to the SiBOC glassy phase [17]. The ceramic conversion has not affected the amorphous SiBOC phase. The IR spectra of the ceramics obtained at 1000°C clearly lack the absorption at 1272 cm^{-1} corresponding to $\nu Si\text{-}CH_3$ supporting the completion of inorganic transformation of the precursor. The Si-O-Si and B-O-Si absorptions are still present in the ceramic as evidenced by the absorptions at 1033 and 887 cm^{-1} as the amorphous phase has not undergone much phase transitions during the pyrolysis. From literature it is well known that siloxanes on pyrolysis give siliconoxycarbide glasses. Compared to the parent silica glass, the presence of tetra coordinated carbon units in SiOC leads to a stiffer and constrained network with improved mechanical properties as well as density, viscosity and glass transition temperature. The carbon containing network has also a lower tendency towards devitrification [18]. SiOC glass phase decomposes in the temperature range 1200-1500°C, leading to the precipitation of nano sized β-SiC [19]. The effect of carbon content plays an important role in these processes [20]. Boron contents of the borosiloxane and the SiBOC ceramic are 3.4 and 4.2 % respectively. The slight increase is attributed to the loss of organic moieties and the corresponding reduction in carbon content during pyrolysis.

The ceramic obtained by the pyrolysis of methylborosiloxane is a boron modified SiOC phase, in which $SiCO_3$, SiC_2O_2, SiC_3O, BC_2O and BCO_2 units may be present and the relative concentration of each unit would depend on the pyrolysis temperature. The formation of SiC nuclei occurs through redistribution reactions of Si-C, B-C, Si-O and B-O bonds that involve, at the same time, mixed silicon oxycarbide (SiC_xO_{4-x}, $0 < x < 4$) and boron oxycarbide (BC_yO_{3-y}, $0 < y < 3$) units. It is reported [6] that the exchange / redistribution reactions take place in the temperature range 1200° to 1400°C. At higher temperatures, the boron can influence the crystallization of SiOC ceramics to SiC[6,7].

Siliconboronoxycarbide foam:

The siliconboronoxycarbide (SiBOC) foam (Fig. 5 a&b) was obtained from borosiloxane coated polyurethane after pyrolysis at 1000°C and 1400°C (Fig. 5 c &d).

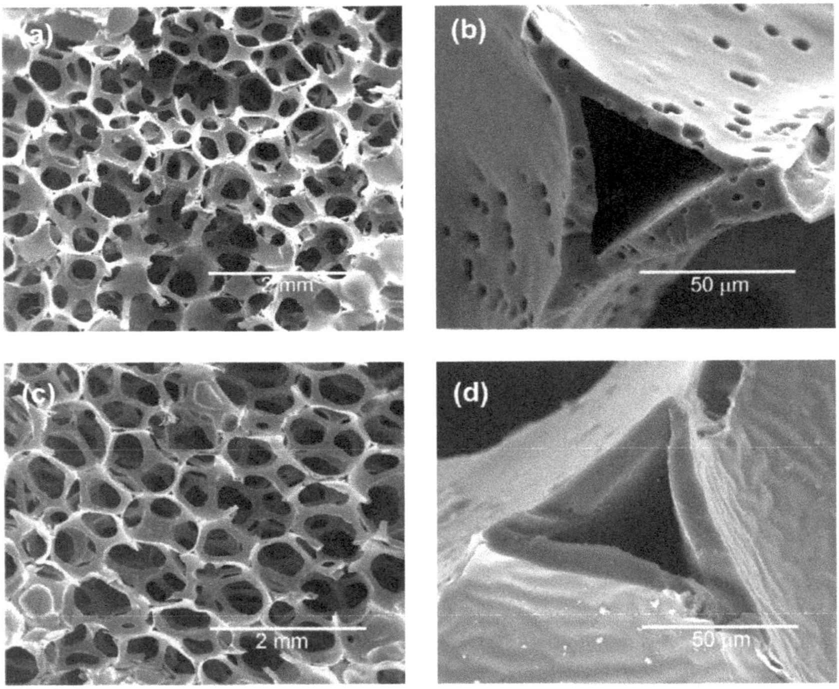

Fig. 5: SEM images of SiBOC foam pyrolyzed at 1000°C(a&b) and 1400°C(c&d).

The foam showed good mechanical integrity and shape retention with shrinkage. The strut density was measured to be 2.11 g/cm^3 at 1000°C and 2.15 g/cm^3 for 1400°C. Geometrical density decreased from 0.15g/ cm^3 (PU-foam) to 0.18g/ cm^3 (overall change 22%) for the ceramic foam. The linear shrinkage varies between 15 up to 19% regarding to measured direction leading to a calculate volume shrinkage of 44%. SiOBC-foams exhibit a different strut microstructure than common SiOC foams. Due to the replica process and removing PU-foam support the ceramics foamstrut crosscut usally shows a triangle shape with partially detached walls (schematic drawing Fig. 6a).

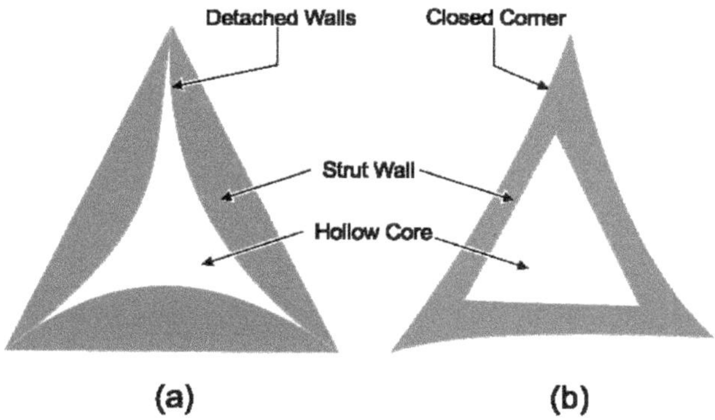

Fig. 6. Schematic drawing of cross sectional view of a) SiOC and b) SiBOC foam strut.

But the SiBOC foam instead showed no detached walls and sharp edges (Fig. 6b) with closed corner gaps (Fig. 5 d). Due to the lower high temperature viscosity of SiBOC glass compared to that of SiOC, made possible by the presence of boron can the low viscous matrix partially melt and flow. Thereby the gaps and corner edges are closed and the walls of the strut joined together.

Another important feature of the foam material was the presence of a secondary porosity on the strut walls (Fig. 5 b). The SEM analysis of the precursor foam after cross linking at 125°C showed no evidences of this porosity. This observation confirms that these pores are not formed by the evaporation of ethanol from the precursor. On the other hand, the formation of these pores has happened during the ceramic conversion of the precursor foam. This secondary porosity is also reduced by pyrolysing at T_{pyro} =1400°C due to viscous matrix flow. This is proofed by the μCT-results where primary the amount of small cells sizes (representing second porosity) is reduced and secondary the cells slightly expand by increasing mean cell diameter 656 μm up to 756 μm (Fig. 7a).

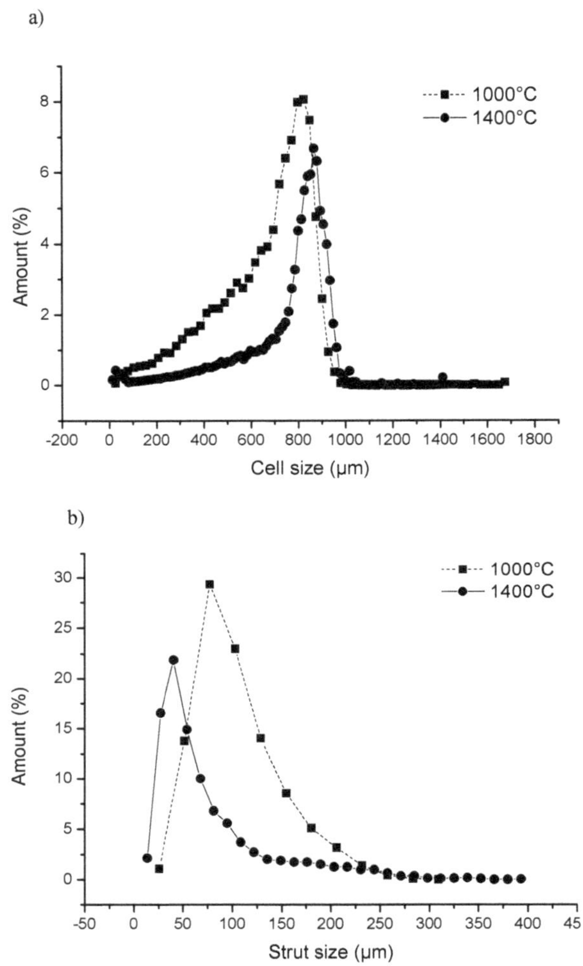

Fig. 7. Distribution of cell size (a) and strut size (b) for 1000°C and 1400°C.

The viscous matrix flow without additional volume shrinkage up to T_{pyro} = 1400°C leads to thinner struts (matching expanding cells) as shown in Fig 7b, reducing mean strut thickness from 105 μm (T_{pyro} = 1000°C) to down to 78 μm (T_{pyro} = 1400°C). Also there is no additional volume shrinkage due the major degradations and micro structural change induced dimensional changes are completed at 1000°C.

It is assumed that mass conservation and foam structure evolution are scale invariant that fractional density ρ* of a polyhedral foam was expressed as a function of geometrical variables cell size d and cell wall thickness δ by[21]

$$\rho^* = 1 - \frac{1}{\left(1+\frac{\delta}{d}\right)^3} \qquad (1)$$

Using equation 1 with experimental μ-CT values mentioned above for the foam microstructure the quotient $r^*_{1000°C}$ / $r^*_{1400°C}$ = 1 indicates no volume and porosity change. As discussed earlier, the only major change happening is the SiC crystallization from the amorphous matrix by exchange mechanism between SiOC tetrahedra. The SiOBC foams pyrolyzed at 1000°C showed an average compressive strength value of 420 kPa with an Weibull modulus m = 4.302 and sigma_0 = 0.465 MPa. Due to either cell size expansion and strut thickness reduction at T_{pyro} = 1400°C the compressive strength decreases down to 180 kPA with an Weibull modulus m = 3.690 and sigma_0 = 0.195 MPa. The total volume fraction of preceramic boronsiloxan precursor after curing, $\phi_{Pre}{}^{inf}$, impregnating the PU foam of fractional density ρ^*_{PU} (e.g. the solid volume fraction of raw foam) is given by[24]

$$\phi_{Pre}{}^{\inf} = \frac{\omega_{Pre}}{(1-\omega_{Pre})}\frac{\rho_{PU}}{\rho_{Pre}}\left(\frac{\rho^*_{PU}}{1-\rho^*_{PU}}\right) \qquad (2)$$

where ω_{Pre} is the weight fraction of preceramic boronsiloxan precursor and ρ_{PU} and ρ_{Pre} are the densities of the PU solid (strut) (ρ_{PU} = 1200 kg/m^3) and of the precursor (ρ_{Pre} = 947.2 kg/m^3).

Note that equation (2) is independent on the specific cell structure of the foam and $\phi_{Pre}{}^{inf}$ may vary from 0 (no infiltration) to 1 (foam porosity completely filled with slurry). Thus, for ω_A = 0.764 and ρ^*_{PU} = 0.035 a volume fraction of infiltrated slurry A of $\phi_A{}^{inf}$= 0.1484 s calculated from equ. (1). Approximating the cell structure of the surface coated reticulated foam by tetrakaideca-hedron cells, a mean coating layer thickness of slurry A, c_A, covering the impregnated foam cell surface can be estimated for the case that all impregnated slurry covers the cell struts homogeneously from[22] by equation (3) and (4)

$$c_A = \frac{t_{PU+A} - t_{PU}}{2} \qquad (3)$$

$$\frac{t_{PU+A}}{t_{PU}} \approx \sqrt{\frac{\phi_A{}^{\inf} + \rho^*_{PU}}{\rho^*_{PU}}} \qquad (4)$$

where t_{PU} is the thickness of the PU cell strut and t_{PU+A} is the thickness of the impregnated foam cell strut. Inserting $\phi_A = 0.764$ a $t_{PU+A}/t_{PU} \approx 2.29$ is derived which for $t_{PU} = 70$ µm equals to a mean coating layer thickness c_A of approximately 35 µm (taking into account a shrinkage of the green body of 22%).

CONCLUSIONS

Siliconboronoxycarbide foam was prepared from methylborosiloxane precursor, via the polyurethane template method. The novel borosiloxanes precursor on pyrolysis at 1000°C under inert atmosphere gives amorphous SiBOC ceramics. The SiBOC foams prepared from this precursor showed good shape retention and mechanical integrity with a 44.2% volume shrinkage with respect to the precursor foam. The bulk and strut densities were measured to be 0.18 and 2.11 (1000°C) and 2.15 (1440°C) g/cm^3 respectively. The material showed an average compressive strength of 420 kPa. Further heat treatment of the foam at 1400°C resulted in the formation of a SiC/SiBOC nanocomposite foam. The lower compressive strength of the nano composite foam was attributed to the crystalline matrix and cell size expansion combined with strut thickness thinning.

The authors gratefully acknowledge the funding of the German Research Foundation (DFG), which, within the framework of its "Excellence Initiative", supports the Cluster of Excellence "Engineering of Advanced Materials" (www.eam.uni-erlangen.de) at the University of Erlangen-Nuremberg.

REFERENCES

[1] M. Scheffler and P. Colombo, Eds., *Cellular Ceramics: Structure, Manufacturing, Properties and Applications*. Weinheim: Wiley-VCH Verlag GmbH & Co. KGaA, 2005, p. 645.

[2] P. Colombo, R. Riedel, G. D. Soraru, and H.-J. Kleebe, Eds., *Polymer Derived Ceramics: From Nano-Structure to Applications*. Lancaster: Destech Publications, Inc., 2010, p. 476.

[3] P. Colombo, G. Mera, R. Riedel, and G. D. Sorarù, "Polymer-Derived Ceramics: 40 Years of Research and Innovation in Advanced Ceramics," *Journal of the American Ceramic Society*, vol. 93, no. 7, pp. 1805–1837, Jun. 2010.

[4] G. M. Renlund, S. Prochazka, and R. H. Doremus, "Silicon oxycarbide glasses: Part II. Structure and properties," *Journal of Materials Research*, vol. 6, no. 12, pp. 2723–2734, Jan. 2011.

[5] H. Bréquel, J. Parmentier, S. Walter, R. Badheka, G. Trimmel, S. Masse, J. Latournerie, P. Dempsey, C. Turquat, A. Desmartin-Chomel, L. Le Neindre-Prum, U. A. Jayasooriya, D. Hourlier, H.-J. Kleebe, G. D. Sorarù, S. Enzo, and F. Babonneau, "Systematic Structural Characterization of the High-Temperature Behavior of Nearly Stoichiometric Silicon Oxycarbide Glasses," *Chemistry of Materials*, vol. 16, no. 13, pp. 2585–2598, Jun. 2004.

[6] M. A. Schiavon, C. Gervais, F. Babonneau, and G. D. Soraru, "Crystallization Behavior of Novel Silicon Boron Oxycarbide Glasses," *Journal of the American Ceramic Society*, vol. 87, no. 2, pp. 203–208, Feb. 2004.

[7] D. Devapal, S. Packirisamy, K. J. Sreejith, P. V. Ravindran, and B. K. George, "Synthesis, Characterization and Ceramic Conversion Studies of Borosiloxane Oligomers from Phenyltrialkoxysilanes," *Journal of Inorganic and Organometallic Polymers and Materials*, vol. 20, no. 4, pp. 666–674, Jun. 2010.

8 R. Riedel, J. Bill, and A. Kienzle, "Boron-modified Inorganic Polymers - Precursors for the Synthesis of Multicomponent Ceramics," *Applied Organometallic Chemistry*, vol. 10, no. 3–4, pp. 241–256, 1996.

9 J. C. Salamone, Ed., *Polymeric Materials Encyclopedia: Polyborosiloxanes, Vol. 7.* New York: CRC Press, 1996.

10 R. Riedel, *Processing of Ceramics, Part II, Volume 17B, Materials Science and Technology: A Comprehensive Treatment.* Weinheim: Wiley-VCH Verlag GmbH & Co. KGaA, 1996.

11 S. Yajima, T. Shishido, and M. Hamano, "SiC and Si3N4 sintered bodies with new borodiphenylsiloxane polymers as binder," *Nature*, vol. 266, no. 5602, pp. 522–524, Apr. 1977.

12 R. Peña-Alonso and G. D. Sorarù, "Synthesis and characterization of hybrid borosiloxane gels as precursors for Si-B-O-C fibers," *Journal of Sol-Gel Science and Technology2*, vol. 43, pp. 313–319, 2007.

13 R. L. Siqueira, I. V. P. Yoshida, L. C. Pardini, and M. A. Schiavon, "Poly(borosiloxanes) as precursors for carbon fiber ceramic matrix composites," *Materials Research*, vol. 10, no. 2, pp. 147–151, 2007.

14 K. J. Sreejith and S. Packirisamy, "Phenyl borosiloxane derived ceramic matrix composites," in *High Temperature Ceramic Materials and Composites*, W. Krenkel and J. Lamon, Eds. Berlin: AVISO Verlagsegesellschaft mbH, 2010, pp. 712–718.

15 A. Karakuscu, R. Guider, L. Pavesi, and G. D. Sorarù, "Broad-band tunable visible emission of sol–gel derived SiBOC ceramic thin films," *Thin Solid Films*, vol. 519, no. 11, pp. 3822–3826, Mar. 2011.

16 K. Schwartzwalder and A. V. Sommers, "Medthod of making porous ceramic articles," U.S. Patent 30900941963.

17 G. D. Soraru, F. Babonneau, S. Maurina, and J. Vicens, "Sol-gel synthesis of SiBOC glasses," *Journal of Non-Crystalline Solids*, vol. 224, no. 2, pp. 173–183, Mar. 1998.

18 A. D. Chomel, P. Dempsey, J. Latournerie, D. Hourlier-Bahloul, and U. A. Jayasooriya, "Gel to Glass Transformation of Methyltriethoxysilane: A Silicon Oxycarbide Glass Precursor Investigated Using Vibrational Spectroscopy," *Chemistry of Materials*, vol. 17, no. 17, pp. 4468–4473, 2005.

19 F. Babonneau and J. Maquet, "Nuclear magnetic resonance techniques for the structural characterization of siloxane – oxide hybrid materials," *Polyhedron*, vol. 19, pp. 315–322, 2000.

20 J. Cordelair and P. Greil, "Electrical conductivity measurements as a microprobe for structure transitions in polysiloxane derived Si – O – C ceramics," *Journal of the European Ceramic Society*, vol. 20, no. 12, pp. 1947–1957, 2000.

21 M. Scheffler, T. Gambaryan-Roisman, T. Takahashi, J. Kaschta, H. Muenstedt, P. Buhler, P. Greil, *Ceram.Trans.* 2000, *115*, 239-250

22 L. J. Gibson, M.F. Ashby, *Cellular Solids*, Cambridge Univ. Press, 2nd edition 1997

23 B. Ceron-Nicolat, F. Wolff, A. Dakkouri-Baldauf, T. Fey,* H. Münstedt, P. Greil: *Graded Cellular Ceramics from Continuous Foam Extrusion*, Adv. Eng. Materials, 2012,DOI: 10.1002/adem.201200039

[24] B. Ceron-Nicolat, T. Fey, P. Greil: *Processing of Ceramic Foams with Hierarchical Cell Structure*, Adv. Eng. Materials, 2010, 884-892, DOI: 10.1002/adem.201000114

SYNTHESIS OF A POROUS SiC MATERIAL FROM POLYCARBOSILANE BY DIRECT FOAMING AND RADIATION CURING

Akira Idesaki, Masaki Sugimoto and Masahito Yoshikawa
Japan Atomic Energy Agency
Takasaki, Gunma, Japan

ABSTRACT

A macroporous SiC material was synthesized from polycarbosilane (PCS) by direct foaming and radiation curing. A mixture of PCS and biurea, a blowing agent, was heat-treated in Ar to obtain foamed PCS body, irradiated by gamma-rays to fix the shaped body and finally pyrolyzed at 1000°C in Ar. In this work, effects of foaming temperature (280-320°C) and biurea content (0-1wt%) on the properties of obtained porous SiC material such as average cell size, porosity, compressive strength were investigated. It was found that the average cell size increased as foaming temperature and/or biurea content increased. Corresponding to increasing in average cell size, total porosity and open porosity increased and bulk density, closed porosity and compressive strength decreased. In case of the foaming at 280°C with biurea content of 1wt%, obtained porous SiC material showed average cell size of 0.5mm, total porosity of 74% and compressive strength of 3.2MPa.

INTRODUCTION

Porous ceramic materials which possess low density, high specific surface area (SSA), high gas/liquid permeability and low thermal conductivity as well as intrinsic properties of ceramic materials are utilized as catalyst supports, hot gas filters, molten metal filters, thermal insulation, and so on, in wide industrial fields [1-3]. In order to expand these applications more, it is important to develop porous ceramic materials with enhanced properties such as thermal resistance, corrosive resistance, strength, and so on. Silicon carbide (SiC) which possesses excellent resistance for heat, oxidation and corrosion can give porous ceramic materials available under severe conditions. Several fabrication processes of porous SiC materials have been proposed: "sacrificial template method" using polymeric spheres, polymeric foams and freeze-casting of slurry or solution and "direct foaming method" using blowing agents or CO_2 gas [4-9]. While SiC ceramic powders and preceramic polymers are used in these processes, the fabrication process using preceramic polymers is advantageous in comparison with powder-sintering process because the preceramic polymers possess high formability, low processing temperature and controllability of compositions with addition of fillers. We have synthesized a variety of SiC products such as fibers, micro parts, membranes for gas separation from preceramic polymers with radiation curing [10-14]. In this work, fabrication of a porous SiC material from polycarbosilane (PCS) was examined with following processes (Fig. 1): (1) direct foaming process; PCS is heated to melt and simultaneously blowing agent is thermally decomposed to generate gases for foaming, (2) radiation curing process; the foamed body is irradiated by gamma-rays in air to fix the foamed shape and (3) pyrolysis process; the cured body is converted to a porous SiC material by firing at 1000°C in an inert atmosphere. Because an optimum blowing agent for PCS should be selected to obtain a porous SiC material, biurea which decomposes in a temperature range of 180-340°C was used in this work. In order to develop SiC products of catalyst support, hot gas filter, molten metal filter, and so on, it is necessary to optimize the properties of porous SiC material such as pore size distribution, porosity, SSA, strength and permeability; small pores with nanometer size, high open porosity, high SSA and corrosive resistance are required particularly for catalyst supports, high open porosity, high strength, high permeability and thermal resistance for filters. Therefore, effects of foaming temperature and biurea content on the properties of obtained porous SiC material were investigated.

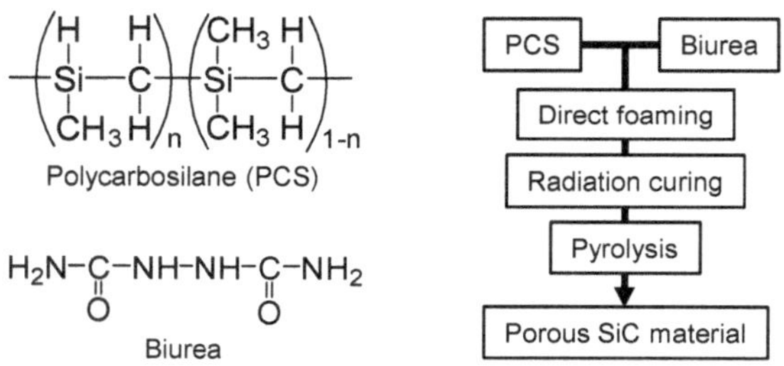

Figure 1. Chemical structures of starting materials and fabrication process of a porous SiC material in this work.

EXPERIMENTAL PROCEDURE

Polycarbosilane (PCS, NIPUSI Type A, Nippon Carbon Co., Ltd.) is a solid material at ambient temperature with the number average molecular weight and density of 2000g/mol and $1.1 g/cm^3$, respectively. Biurea (Tokyo Chemicals Industry Co., Ltd.) is a white powder which decomposes above 180°C evolving NH_3, CO_2 and N_2 with the amount of 90ml/g [15]. A mixture of PCS and biurea put into an aluminum pan was heated to 230°C with a heating rate of 200°C/h and hold for 2h, and then heated to a foaming temperature with the heating rate of 200°C/h under Ar atmosphere. The sample was not held at the foaming temperature and cooled in the furnace to obtain a foamed PCS body. The foamed PCS body was irradiated by ^{60}Co gamma-rays with dose of 1MGy at room temperature in air. After the irradiation, the foamed PCS body was heat-treated at 250°C for 1h in Ar to complete curing process. The cured PCS body was pyrolyzed at 1000°C for 1h in Ar.

In order to determine the foaming temperature, measurement of melt viscosity of PCS and thermogravimetric analysis of PCS and biurea were conducted under an inert atmosphere. Obtained porous SiC materials were observed by scanning electron microscopy (SEM). Average cell size was calculated from SEM images using the ImageTool software (UTHSCSA, University of Texas, San Antonio, TX, USA). The raw data obtained by image analysis were converted to 3D values to obtain the effective cell dimension by applying the stereological equation (1) [16]:

$$D_{sphere} = D_{circle}/0.785 \qquad (1)$$

Bulk density and porosity of obtained porous SiC material were measured by Archimedes method using cyclohexane as a buoyant. Compressive test was conducted for three cubic specimens with size of 5×5×5mm at a loading speed of 0.5mm/min.

RESULTS AND DISCUSSION

In the direct foaming process, PCS is heated to melt and simultaneously biurea is thermally decomposed to generate gases for the foaming. PCS needs to be heated in an inert atmosphere in order to prevent oxidation. Figure 2 shows the results of change in melt viscosity of PCS in Ar (Fig. 2(a)) and thermogravimetric (TG) curves of PCS and biurea in He (Fig. 2(b)). PCS melted above 220°C and its viscosity decreased drastically with increasing in temperature; the viscosity was 77, 19 and 5Pa·s at 280, 300 and 320°C, respectively. In the temperature range up to 400°C, mass loss of PCS was almost negligible. On the other hand, thermal decomposition of biurea started above 180°C, and TG curve dropped drastically up to 340°C; the mass residue was 64, 46 and 18% at 280, 300 and 320°C, respectively. If all of the evaporating components are assumed as gas components, the amount of generated gas from biurea is estimated as 32, 49

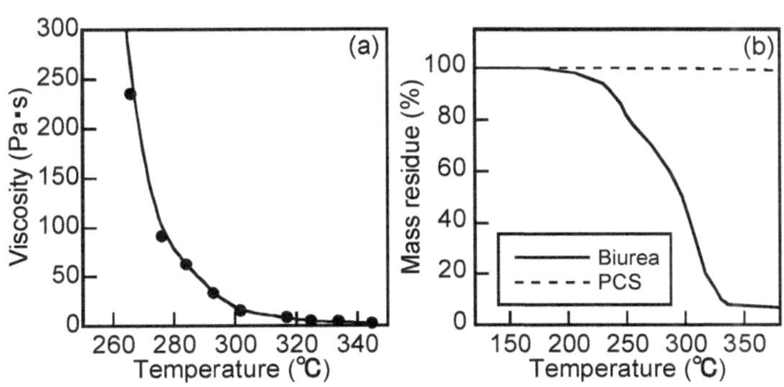

Figure 2. (a) Change in melt viscosity of PCS in Ar and (b) Thermogravimetric curves of PCS and biurea in He.

and 74ml/g at 280, 300 and 320°C, respectively. According to the results above, the foaming temperature was set at 280, 300 and 320°C in order to examine the effect of foaming temperature on properties of obtained porous SiC material under conditions of varying viscosity of PCS and gas volume from biurea; 280°C: high viscosity and small gas volume, 300°C: intermediate viscosity and intermediate gas volume, and 320°C: low viscosity and large gas volume.

Figure 3 shows expansion rate of PCS after the foaming at 280°C. The expansion rate was calculated by following equation (2):

$$\text{Expansion rate (\%)} = \frac{\text{Height after foaming (mm)}}{\text{Height at melt (mm)}} \times 100 \qquad (2)$$

where height at melt was calculated by weight and density ($1.1 g/cm^3$) of PCS and diameter of aluminum pan. The result indicates that pure PCS can be self-foamed with expansion rate of about 240%. Such self-foaming is considered to be caused by dissolved air in PCS because the mass decrease was almost negligible at 280°C (Fig. 2(b)). The expansion rate of PCS increased with increasing in biurea content, and it reached to about 600% with biurea content of 1wt%. In case of biurea content of more than 1wt%, very large cells with diameter of a few millimeters were observed. Such large cells make the foamed body too brittle to handle and influence on mechanical property of obtained porous SiC material, therefore, biurea content of 0.5 and 1wt% was adopted for the following experiments in this work.

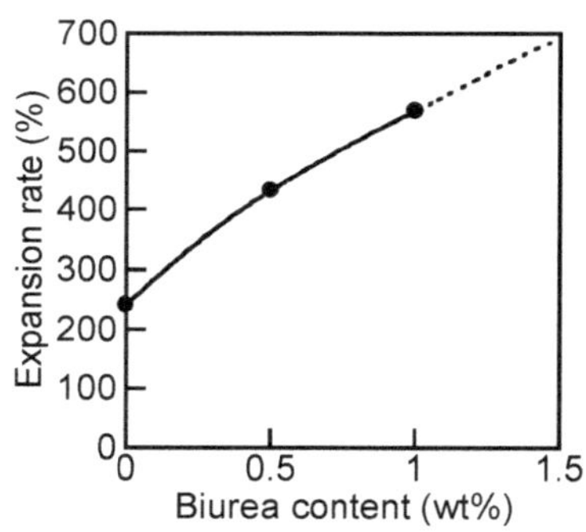

Figure 3. Expansion rate of PCS after foaming at 280°C with different biurea content.

After the foaming process, foamed PCS body was cured by gamma-ray irradiation in air and subsequent heat treatment in Ar to fix the foamed shape. Irradiation of gamma-rays is advantageous in the view point of penetration, that is, thickness of the sample. The half-value thickness of ^{60}Co gamma-rays in water is about 11cm where intensity of gamma-rays decays to half [17]. In case of PCS, the half-value thickness is estimated as about 10cm according to its density. Practically, a sample with thickness of 10cm towards the gamma-ray source can be irradiated homogeneously by turning back and forth at a half of time for the irradiation. According to our previous work, PCS is cured sufficiently by gamma-ray irradiation with dose of about 1MGy in case of radiation oxidation curing [10]. If the oxidation is not sufficient, PCS melts during the pyrolysis process. Figure 4 shows SEM images of porous SiC materials obtained under various conditions. In case of pure PCS (Fig.4 (a)), melting of PCS was observed as wrinkled part in left side of image. This is caused by insufficient oxidation during gamma-ray

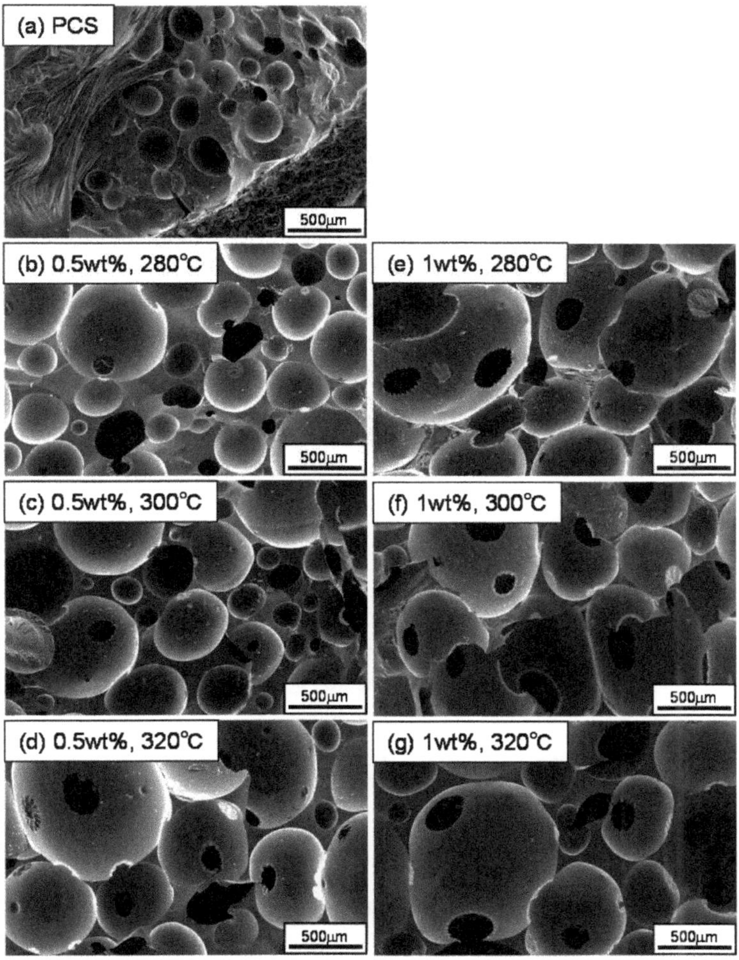

Figure 4. SEM images of porous SiC materials obtained under various conditions.

irradiation due to presence of many closed pores. By adding biurea, open pores are obtained, that leads sufficient penetration of oxygen. Even biurea content of 0.5wt% is effective to synthesize a porous SiC material with high interconnectivity. Comparing images from figure 4(b) to 4(g), cell size of the porous SiC material becomes large as biurea content or foaming temperature increase. This is caused by increased gas volume from biurea due to high biurea content or high foaming temperature. After the pyrolysis process, foamed PCS body shrunk by about 30% in linear. The shrinkage was three-dimensionally homogeneous, and there was no large crack in the sample. Figure 5 (a) shows average cell size of obtained porous SiC material. Cell size depends on balance between gas volume from biurea and viscosity of PCS. Cell size becomes large under a condition of large gas volume and constant viscosity, that is, with increasing in biurea content at constant foaming temperature. On the other hand, under a condition of constant gas volume (constant biurea content), cell size is hard to be large with high viscosity (low temperature) and is easy to be large with low viscosity (high temperature). These support the difference in average cell size of obtained porous SiC materials with different conditions. According to this, cell size becomes the largest under a condition of large gas volume (high biurea content) and low viscosity (high foaming temperature), however, the obtained average cell size saturated at around 0.5mm. This is considered to be caused by escaping of gases which do not attribute to the foaming of PCS. If cell size is constant and evolved gases do not escape out of PCS, the expansion rate after foaming should increase with increasing in the gas volume from biurea. However, as shown in Fig. 5(b), the expansion rate was almost same between 300 and 320°C for the sample with biurea content of 1wt%. This result supports the presence of gases which escape out of PCS.

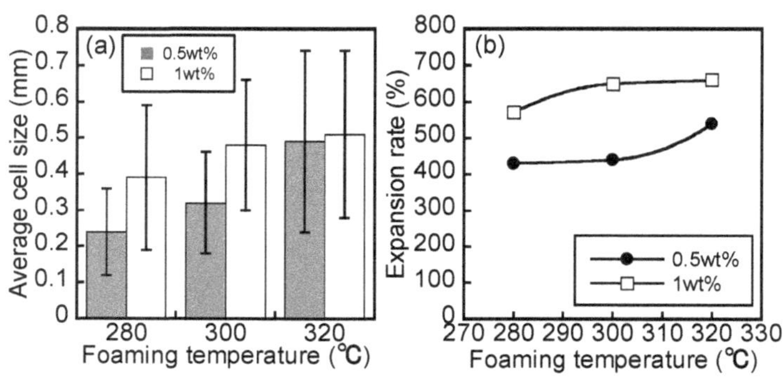

Figure 5. (a) average cell size of obtained porous SiC materials and (b) relationship between expansion rate of PCS (before pyrolysis) and foaming temperature.

Cell size and expansion rate influence on bulk density and total porosity of obtained porous SiC material. Figure 6(a) shows that the bulk density decreases and the total porosity increases with increasing in cell size and expansion rate. Under the condition of biurea content of 1wt% and foaming temperature of 280°C, bulk density of 0.677g/cm^3 and total porosity of 74% were obtained. Furthermore, increasing in cell size makes "membrane" formed between adjacent cells thin, that causes rupture of membrane in the pyrolysis process to increase in the open porosity and to decrease in the closed porosity correspondingly. Under the condition of biurea content of 1wt% and foaming temperature of 280°C, open porosity of 65% and closed porosity of 9% were obtained (Fig. 6(b)). On the other hand, increasing in total porosity, open porosity and cell size generates thin cell walls which cause decreasing in compressive strength of obtained porous SiC material. The compressive strength of obtained porous SiC material decreased with increasing in open porosity and cell size for both material obtained with biurea content of 0.5 and 1wt% (Fig. 6(c)). The porous SiC material obtained with biurea content of 1wt% and foaming temperature of 280°C showed the highest average compressive strength of 3.2MPa.

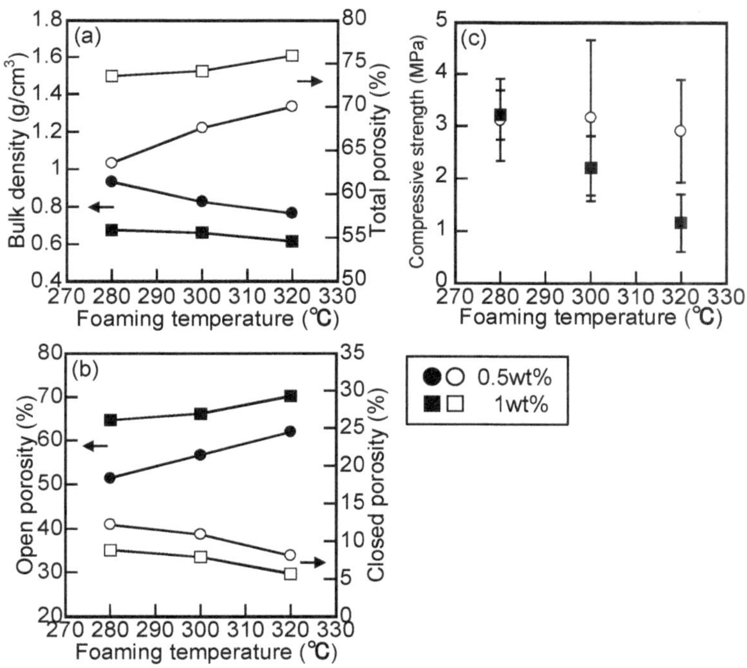

Figure 6. Properties of obtained porous SiC materials: (a) bulk density and total porosity, (b) open porosity and closed porosity and (c) compressive strength.

CONCLUSION

In this work, a macroporous SiC material was synthesized from polycarbosilane (PCS) by direct foaming and radiation curing. Biurea was selected as a blowing agent for PCS, and it was found that biurea is very useful for the foaming because its thermal decomposition temperature range matches with rheological behavior of PCS. PCS was foamed by heat treatment with biurea in Ar, cured by gamma-ray irradiation in air and subsequent heat treatment in Ar and finally pyrolyzed at 1000°C in Ar. In case of the foaming at 280°C with biurea content of 1wt%, obtained porous SiC material showed an average cell size of 0.5mm, total porosity of 74% and compressive strength of 3.2MPa. A porous SiC material which possesses high open porosity and high compressive strength was synthesized from PCS and biurea. This material is expected to be applied for a catalyst support, a hot gas filter or a molten metal filter although evaluation and optimization of specific surface area and gas/liquid permeability are required.

REFERENCES

[1] M. Scheffler and P. Colombo, Cellular ceramics, Wiley-VCH Verlag GmbH, Weinheim, (2006).

[2] A. R. Studart, U. T. Gonzenbach, E. Tervoort, and L. J. Gauckler, Processing Routes to Macroporous Ceramics: A Review, *J. Am. Ceram. Soc.*, **89**[6], 1771–1789 (2006).

[3] P. Colombo, C. Vakifahmetoglu and S. Costacurta, Fabrication of ceramic components with hierarchical porosity, *J. Mater. Sci.*, **45**, 5425-5455 (2010).

[4] Y. W. Kim, Y. J. Jin, J. H. Eom, I. H. Song and H. D. Kim, Engineering porosity in silicon carbide ceramics, *J. Mater. Sci.*, **45**[10], 2808-2815 (2010).

[5] X. Zhu, D. Jiang, S. Tan, Preparation of silicon carbide reticulated porous ceramics, *Mater. Sci. Eng.: A*, **323**, 232–238 (2002).

[6] M. Fukushima, M. Nakata, Y. Zhou, T. Ohji and Y. Yoshizawa, Fabrication and properties of ultra highly porous silicon carbide by the gelation–freezing method, *J. Europ. Ceram. Soc.*, **30**[14], 2889–2896 (2010).

[7] B. H. Yoon, E. J. Lee, H. E. Kim and Y. H. Koh, Highly aligned porous silicon carbide ceramics by freezing polycarbosilane/camphene solution, *J. Am. Ceram. Soc.*, **90**[6], 1753–1759 (2007).

[8] M. Fukushima and P. Colombo, Silicon carbide-based foams from direct blowing of polycarbosilane, *J. Europ. Ceram. Soc.*, **32**[2], 503–510 (2012).

[9] Y. W. Kim and C. B. Park, Processing of microcellular preceramics using carbon dioxide, *Comp. Sci. Tech.*, **63**, 2371–2377 (2003).

[10] M. Narisawa, A. Idesaki, S. Kitano, K. Okamura, M. Sugimoto, T. Seguchi and M. Itoh, Use of blended precursors of polyvinylsilane in polycarbosilane for silicon carbide fiber synthesis with radiation curing, *J. Am. Ceram. Soc.*, **82**[4], 1045-1051 (1999).

[11] A. Idesaki, M. Narisawa, K. Okamura, M. Sugimoto, Y. Morita, T. Seguchi and M. Itoh, Fine silicon carbide fibers synthesized from polycarbosilane - polyvinylsilane polymer blend using electron beam curing, *J. Mater. Sci.*, **36**, 357-362 (2001).
[12] A. Idesaki, M. Sugimoto, S. Tanaka, M. Narisawa, K. Okamura and M. Itoh, Synthesis of a Minute SiC Product from Polyvinylsilane with Radiation Curing: Part II -Ceramization Process of Radiation Cured Polyvinylsilane-, *J. Mater. Sci.*, **42**, 130-135 (2007).
[13] R. A. Wach, M. Sugimoto, A. Idesaki and M. Yoshikawa, Molecular sieve SiC-based membrane produced by radiation curing of preceramic polymers for hydrogen separation, *Mater. Sci. Eng.: B*, **140**, 81-89 (2007).
[14] A. Takeyama, M. Sugimoto and M. Yoshikawa, Gas permeation property of SiC membrane using curing of polymer precursor film by electron beam irradiation in helium atmosphere, *Mater. Trans.*, **52**, 1276-1280 (2011).
[15] P. R. Russel and A. N. Strachan, The thermal decomposition of biurea, *J. Chem. Soc., Perkin Trans.*, **2**, 323-326 (1978).
[16] ASTM D 3576. "Standard test method for cell size of rigid cellular plastics"; Annual Book of ASTM Standards, Vol.08.02., West Conshohocken, PA, (1997).
[17] J. W. T. Spinks and R. J. Woods, An introduction to radiation chemistry, John Wiley and Sons Inc., N.Y., (1964).

FABRICATION OF SiOC/C COATINGS ON STAINLESS STEEL USING POLY(PHENYL CARBOSILANE) AND THEIR ANTI-CORROSION PROPERTIES

Yoon Joo Lee, Jong Il Kim, Soo Ryong Kim, Woo Teck Kwon, Dong-Geun Shin, Yonghee Kim[†]

Korea Institute of Ceramic Engineering and Technology
77 Digital-ro, 10-gil, Geumcheon-gu
Seoul 153-801, Korea
[†] E-mail: yhkokim@kicet.re.kr

ABSTRACT

Poly(phenyl carbosilane) (PPCS) has been applied to develop advanced carbon-rich SiOC (SiOC/C) coatings. In this study, STS 316 stainless steel was coated with SiOC/C films, and the chemical structure and anti-corrosion properties of the films were analyzed. The films were obtained at 800 °C under a nitrogen atmosphere by dip coating. Spectroscopic analysis revealed the films to contain high amounts of Si-C bonds in the SiOC networks and free carbon was found to be present with a graphite-like structure. The films showed good adhesion to the substrate and exhibited excellent anti-corrosion properties, especially in HCl solution.

INTRODUCTION

Silicon oxycarbide (SiOC) is an interesting material because of its meta stable properties, excellent biostability, and higher mechanical properties when compared to silicon oxide glass.[1-6] Moreover, SiOC can potentially be used as anti-corrosion films,[7-9] and metals that can be used in extreme conditions have been most sought after industrially.

SiOC films can be prepared by sol-gel method,[5,10-12] chemical vapor deposition (CVD),[13,14] radio frequency magnetron sputtering,[15,16] or by pyrolysis of a silicon-based polymer.[17-20] Simplest methods involving solution processing include dip coating, spin coating, or spray coating that use silicon-based polymer solutions. Poly(phenyl carbosilane) (PPCS, Figure 1), a preceramic polymer soluble in organic solvents, can be used to make films by solution processing. PPCS was originally developed as a SiC precursor; however, the polymer has been studied as a SiOC coating material based on its oxidation properties.[21-24] Moreover, PPCS can supply abundant carbon because of the presence of the phenyl groups and it can be converted into carbon rich SiOC (SiOC/C). The carbon, which exists freely without bonding with the silicon atom, can be converted into glassy carbon even it was prepared at < 1000°C. Such a carbon structure is expected to enhance the anti-corrosion properties and the mechanical properties of SiOC/C.

In previous studies, coatings of SiOC/C films on stainless steel, which were prepared from PPCS solution, showed good anti-corrosion properties in 10% HCl solution at 80 °C.[9] However, the chemical structure of SiOC/C is yet to be elucidated. In this study, SiOC/C films were prepared on stainless steel substrates by dip coating. Further, the chemical structure of the films and the interface between the film and substrate were characterized.

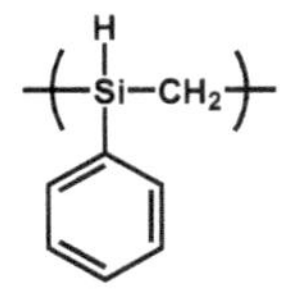

Figure 1. Unit structure of poly(phenyl carbosilane)

EXPERIMENTAL

Coating process

STS 316 stainless steel was chosen as the substrate. The stainless steel was heated at 800 °C in air to form an oxide layer before the coating process. PPCS (TBM Tech, Korea) with a molecular weight of 4,000 was used as the coating material and a 20% solution in toluene was used for the coating.

The SiOC/C films were obtained in three steps: coating, curing, and heat treatment. The dip-coating method was chosen for coating the PPCS and the substrate was pulled out from the coating solution at a speed of 2 mm/min after immersion for 5 min. The coated samples were dried at room temperature overnight and cured at 150 °C in air for 1 h. Subsequently, the samples were heated at 800 °C under N_2.

Characterization of the films

Analysis by powder X-ray diffraction (XRD) was carried out using an X-ray diffractometer (P/MAX 2200V/PC, Rigaku Corp.) with a Cu target (K_α = 1.54 Å). Microscopic images were obtained by scanning electron microscopy (JSM-6700F, JEOL) and atomic force microscopy (JSPM5200, JEOL).

Visible Raman scattering measurements were performed with a JASCO NRS-3100 instrument (Jasco Inc.) using a high-performance 532 nm laser. Solid-state nuclear magnetic resonance (NMR) spectra were recorded (Unity Inova, Varian Associates) at resonance frequencies of 119.182 MHz for ^{29}Si and 150.877 MHz for ^{13}C. A zirconia rotor 5 mm in diameter was used. The magic angle spinning (MAS) technique was used for ^{29}Si at a spinning rate of 9 kHz with a pulse length of 2.5 μs. ^{13}C spectra were measured using the cross-polarization magic angle spinning (CP-MAS) technique at a spinning rate of 8 kHz with a pulse length of 5 μs for 3000 μs.

Corrosion testing

The corrosion tests were performed in a 10% HCl solution. The corrosion test cell was designed as shown in Figure 2. The solution chamber was filled with 10% HCl solution and the test blocks were installed in the four side windows. Then, the solution was heated to 60 °C, and the corrosion test was performed for 24 h.

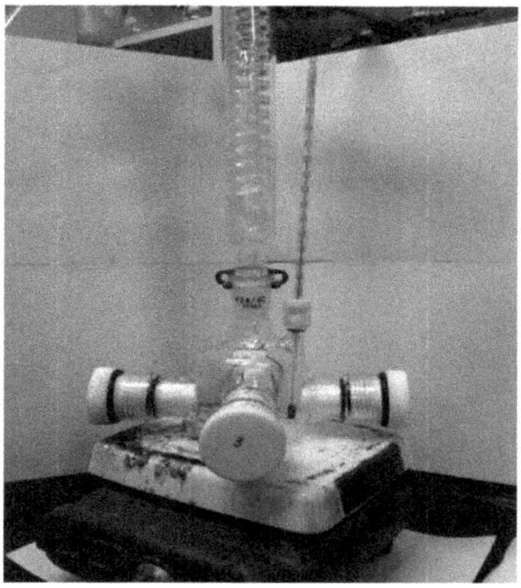

Figure 2. Corrosion test cell with four windows

RESULTS AND DISCUSSION

Chemical structure of SiOC/C

NMR spectroscopy is a powerful method for elucidating chemical structures; however, this technique is inappropriate for analyzing films. In this study, to evaluate the structural properties, we obtained the NMR spectra using flakes. These flakes were prepared by drying the PPCS solution as thin layers in a tray. Then, the SiOC/C flakes were obtained by a method identical to that used in the coating process. The NMR spectra of SiOC/C are displayed in Figure 3. Broad and weak signals were obtained in both ^{29}Si-MAS NMR and ^{13}C CP-MAS NMR spectra because of the loss of protons after the decomposition of the alkyl groups.[25,26]

The ^{29}Si-MAS NMR spectrum showed a broad area from 20 to -120 ppm. The peaks were split into chemical shifts at 20 ppm, -40 ppm, -65 ppm, and -106.5 ppm and each chemical shift was identified with SiO_xC_{4-x} tetrahedral sites (Figure 3 (a)). Despite the lack of clarity in the signal, the ^{29}Si-MAS spectrum indicated that the SiOC/C film, which was prepared at 800 °C using PPCS, contained a large number of Si-C bonds and showed a large area of chemical shift from 0 to -30 ppm corresponding to $SiOC_3$ (near 0 ppm), SiC_4 (-3 ppm), and SiO_2C_2 (-25ppm) states.[27-31]

The ^{13}C CP-MAS spectrum (Figure 3 (b)) showed a single broad peak at 110–160 ppm, corresponding to the presence of sp^2 carbon, while signatures of sp^3 carbon, which can be detected in the 30–50 ppm area[32-34], were found to be absent. In general, amorphous carbon can have two possible states - sp^3 and sp^2. When the amorphous carbon has a greater number of sp^3 centers than sp^2 centers, it is classified as diamond carbon, and the amorphous carbon state composed only of sp^2 carbon is classified as glassy carbon.[35-37] Therefore, the SiOC/C prepared in this study was confirmed to contain free carbon in the form of glassy carbon.

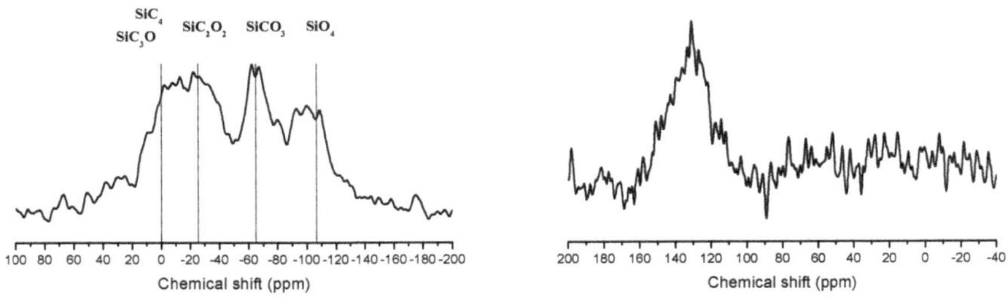

(a) (b)

Figure 3. (a) ^{29}Si MAS and (b) ^{13}C CP-MAS NMR spectra of SiOC/C

In general, sp^2 carbons form a graphitic structure, which can be evaluated by Raman spectroscopy. Figure 4 shows the Raman spectra of the SiOC/C films on stainless steel. Two broad peaks could be observed at 1,580 cm^{-1} and 1,357 cm^{-1}. These two peaks correspond to the carbon modes of the graphitic structure, which shows two main Raman peaks at 1,580 and 1,357 cm^{-1}. The peak at 1,580 cm^{-1} (the G band) is caused by the E_{2g} symmetry and the peak at 1,357 cm^{-1} (D band) is caused by the A_1 symmetry.[38-43] The peak positions and widths of the G and D bands are related to the size and ordering of the graphite carbon cluster. The glassy carbon state or graphitization can be assessed by the peak-intensity ratio of the D and G bands ($I(D)/I(G)$),[44-46] and the value of $I(D)/I(G)$ for the samples considered in the present study was 1.18. From this value, graphitic carbon 30–60 Å in size was estimated to be dispersed in the SiOC/C structure.

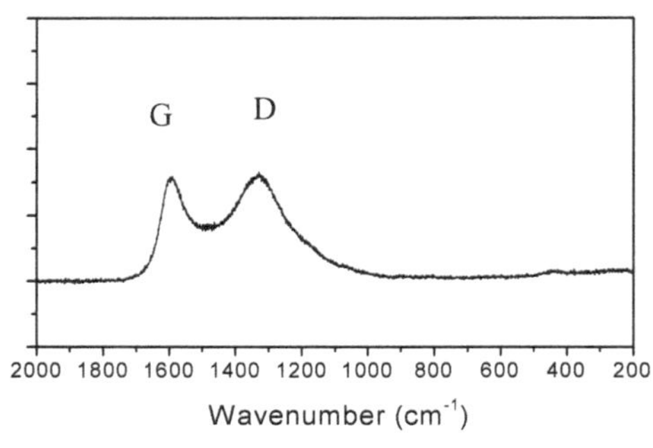

Figure 4. Raman spectrum of SiOC/C film

Oxide layer between the SiOC/C film and the substrate

For applying a ceramic coating on a metal, a bond coat is needed to improve the adhesion between the coating layer and the surface of the substrate. The stainless steel substrate was heated in air before the coating process and the oxide layer, which was formed on the surface of the stainless steel, was used as the bond coat. The surface oxidation was confirmed by

XRD analysis carried out in the thin film mode. From the XRD pattern in figure 5 (b), oxide peaks newly appeared at 29.5°, 33°, and 35°, which corresponded to the presence of $Cr_xFe_{2-x}O_3$. The formation of an oxide layer was confirmed with energy dispersive spectroscopic (EDS) analysis (Table 1).

After the coating process, a glassy black SiOC/C film was formed, which did not peel off when tape testing was performed. Figure 5 (c) shows the sample with the SiOC/C coating. The coating process did not lead to any differences in the XRD patterns; however, the peaks, which were observed after oxidation, became stronger, and a new minute peak was observed at 18°. This new peak was estimated to correspond to an oxide peak, which went undetected in the prior step.

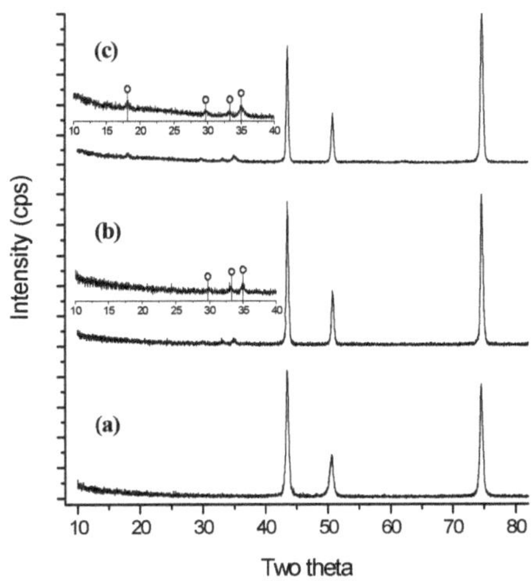

Figure 5. XRD patterns of the surface of (a) the as supplied STS 316 stainless steel, (b) the surface after heat treatment in air, and (c) after the coating process. The peaks related to the oxide structure are marked with circles.

Microscopic analysis of the interface between the SiOC/C film and the substrate

From the microscopic analysis of the sample cross section, the thickness of the SiOC/C film was found to be 900–800 nm. A low resolution TEM image is presented in figure 6 (a). A dense film without empty spaces between the film and the substrate was observed. However, the shape of the interface was nonuniform. Subsequently, EDS analysis performed on the 7 points marked in figure 6 (a) and the atomic composition is listed in table 1. Detection areas 1 and 7 corresponded to the SiOC/C film and the substrate, respectively, while the points 2–6 were at the interface. At the interface, iron and chromium were detected along with silicon and oxygen. Considering the spatial resolution of EDS analysis and the fact that the detection spot size was about ~20 nm (reference of JEOL), it could be confirmed that the nonuniform metal oxide layer was formed on the substrate with a thickness of 100–200 nm. Hence, it acted effectively as a bond coat, showing good adhesion.

Figure 6 (b) shows the high-resolution TEM image of the SiOC/C film. Recalling that Raman spectroscopy indicated a graphite-like structure for the free carbon, the graphitic structure was not clearly observed by TEM because the graphitic carbon cluster was excessively small to be detected by TEM analysis.

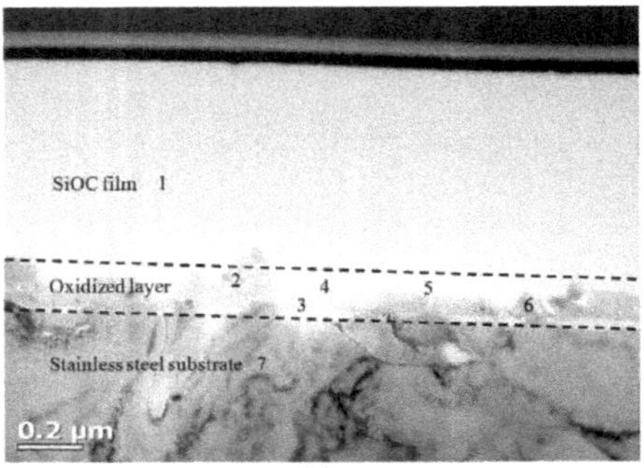

(a)

(b)

Figure 6. (a) Cross-sectional TEM image of the interface between the SiOC/C film and the stainless steel substrate and (b) a high-resolution TEM image of the film layer.

Table 1. Results of the EDS analysis of SiOC/C coated stainless steel samples at points marked in Figure 6.

Atomic percent (%)

Detection area		Si	C	O	Fe	Cr	Ni
SiOC/C film	1	16.78	47.39	35.16	0.39	0.28	-
	2	-	20.50	51.90	2.71	24.88	-
	3	24.79	11.54	57.08	6.58	-	-
Oxidation layer	4	5.23	40.87	35.74	8.25	9.92	-
	5	12.68	57.07	29.34	0.39	0.28	-
	6	-	25.09	42.92	11.21	20.78	-
Stainless steel	7	-	-	-	78.51	13.06	8.79

Corrosion test and microstructure of the test surface

The corrosion test was performed in 10% HCl solution. When the test block was installed in the cell (Figure 2), only a specific area was exposed to the test solution. The test area was 10 mm in diameter. Then, to accelerate the corrosion, the temperature of the test cell was maintained at 60 °C. Figure 7 (a) shows the stainless steel block after the corrosion test was carried out for 24 h. The central circular test area caved in remarkably during the test and could be distinguished clearly from the surface that was not exposed to the test solution. However, hardly any marks were observed on the test surface of the coated stainless steel block, which was tested for 7 days (Figure 7 (b)).

Figure 8 shows the SEM images of the test surfaces of the samples. From the SEM images, it can be seen that the etched surface of the stainless steel was ragged and some corrosion pits were also observed, whereas the surface of the coating was clean. From these results, it was confirmed that the SiOC/C film acted as an excellent protection layer for the stainless steel against HCl solution.

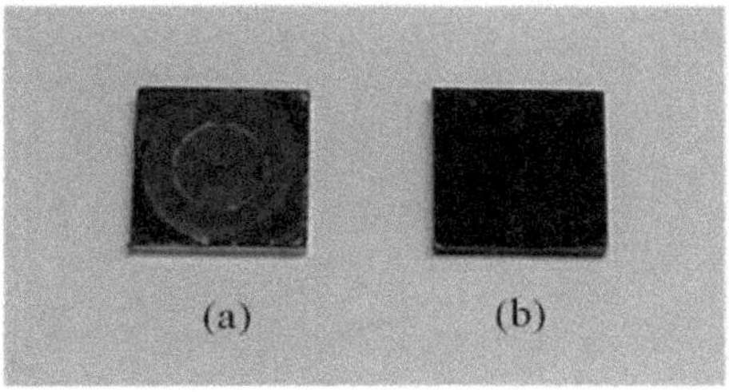

Figure 7. Surfaces of the (a) uncoated sample after corrosion test for 24 h and (b) SiOC/C-coated stainless steel blocks after corrosion test for 1 week in 10% HCl at 60 °C.

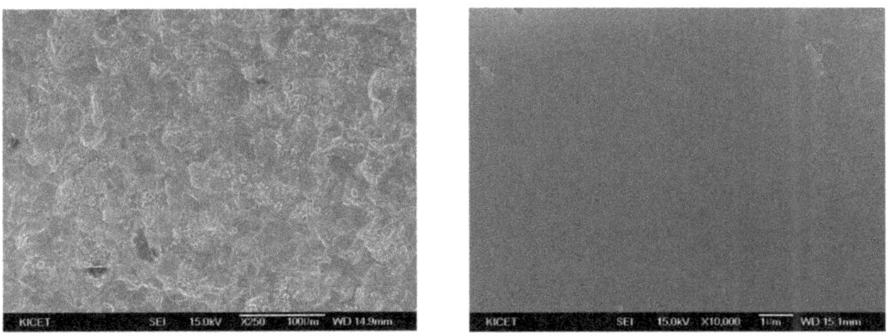

(a) (b)

Figure 8. SEM images of the surface of (a) uncoated stainless steel block after corrosion test for 24 h and (b) SiOC/C-coated stainless steel block after corrosion test for 1 week in 10% HCl solution 60 °C.

CONCLUSIONS

SiOC/C films 700–800 nm in thickness were prepared on stainless steel surfaces by dip coating. The oxide layer at the interface between the coating and the substrate helped in the strong adhesion of the metal to the coating. The SiOC/C film was formed as a dense and uniform glassy structure and defects were absent at the interface of the film and the substrate. The

SiOC/C film derived from PPCS showed excellent corrosion prevention properties, especially in HCl solution. This type of protective coating can be used under extreme conditions.

ACKNOWLEDGMENT

This research was supported by the Pioneer Research Center Program through the National Research Foundation of Korea funded by the Ministry of Science, ICT & Future Planning (NRF-2010-0019468).

PHOTO LUMINESCENT PROPERTIES OF POLYMER DERIVED CERAMICS AT NEAR STOICHIOMETRIC SiO_2-xSiC-y(H) COMPOSITIONS

Masaki Narisawa and Akihiro Iwase
Graduate School of Engineering, Osaka Prefecture University
1-1 Gakuen-Cho, Naka-Ku, Sakai, Osaka 599-8531, Japan
Seiji Watase and Kimihiro Matsukawa
Osaka Municipal Technical Research Institute
1-6-50, Morinomiya, Joto-Ku, Osaka 536-8553, Japan
Taketoshi Kawai
Graduate School of Science, Osaka Prefecture University
1-1 Gakuen-Cho, Naka-Ku, Sakai, Osaka 599-8531, Japan

ABSTRACT

Si-O-C(-H) ceramics were synthesized from silicone resin microspheres with averaged diameters of 2, 6 and 11 μm by using hydrogen for pyrolysis atmosphere. PL and ph osphorescence spectra of samples, in particular, decarbonized at 950 or 1000 °C, were analyzed in detail. During the holding at 950 °C, PL intensity at long wavelength region gradually increased, while the peak positions were at 450 nm. During the holding at 1000°C, increasing rate of the PL intensity at long wavelength region was rapid, and the peak position shifted to 530 nm after 3h holding. Phosphorescence spectra revealed growth of long-lived PL band (580 nm) during the holdings at 950 and 1000 °C, which exactly corresponded to change of the observed PL feature during the holding.

INTRODUCTION

Recently, our group in Osaka Prefecture University has succeeded in synthesizing white Si-O-C(-H) ceramics from densely cross-linked polysiloxane microspheres by pyrolyzing in a hydrogen atmosphere.[1] Such white Si-O-C(-H) ceramics are k nown to exhibit unique photoluminescence (PL).[2-8] In the case of the hydrogen decarbonization, PL appeared at decarbonization temperature of 700-800 °C. At 800-900 °C, Si-O-C(-H) ceramics exhibited blue PL band with a peak positions at 410-420 nm and with a life time of 10-60 ns.[9] Beyond 1000 °C, this blue PL band shifted to longer wavelength with holding the life time of 10-60 ns. On the other hand, a new yellow PL band (peak position of 560-580 nm) appeared beyond 1000 °C. An abnormally long life-time of ca. 2 s was estimated for decay process of this yellow PL. Such long-life times of PL have been rarely observed in Si-O-C ceramics derived from polymer precursors. The old reports about the long-lived PL band of Si-O-C ceramics are, however, isolated, and speculations were not pursued further by other researchers.[10-11] As result of overlapping of the short-lived (blue) and the long-lived (yellow) PL bands in the Si-O-C(-H) ceramics synthesized by the decarbonization reaction, the PL spectra observed at 1000-1300°C became broad, which caused apparent "white" emission. Although the feature of the PL spectrum changed largely at 1000 °C, difference in chemical composition was relatively small ($SiO_{1.54}C_{0.26}H_{0.25}$ (900°C, 1h), $SiO_{1.54}C_{0.26}H_{0.18}$ (1000°C, 3h)). In this article, we s hed light on the growth of the long-lived PL band in the Si-O-C(-H) ceramics, which proceeds at ca. 1000 °C.

EXPERIMENTAL PROCEDURE

Si-O-C(-H) ceramics were synthesized from polysiloxane microspheres in averaged diameters of 2, 6 and 11 μm (Tospearl 120, 2000B and 3120, Momentive Performance Materials Japan) by using hydrogen for the pyrolysis atmosphere. Heating rates and hydrogen flow rates used for the decarbonization were described in previous studies.[1,9] Holding times at 950 and 1000 °C were adjusted in a time range of 10 min - 3h. PL spectra of the Si-O-C(-H) ceramics were observed using an F-4500 (Hitachi, High-Technologies Corporation, Tokyo, Japan). Phosphorescence spectra were obtained by the same F-4500 apparatus by shuttering short-life time PL components below 1 ms. FT-IR spectra were observed by Spectrum GX (PerkinElmer Japan Co., Ltd., Tokyo, Japan) by diffuse reflection method. Specific gravities were measured by a picnometer using pure water as medium.

RESULTS AND DISCUSSION

Figures 1 (a) and (b) show photographs the Si-O-C(H) ceramics with decarbonization temperatures of 950 and 1000 °C under a visible light and a UV light. Ceramic yields from the precursor after decarbonization were always at 78-79%. There is no difference in color of the samples under the visible light irradiation. Under the UV light irradiation, however, the sample decarbonized at 1000 °C for 3h exhibits white emission, while other samples exhibits almost the same blue emission. As expected from a previous study, the long-lived PL, which lasted for a few seconds after stopping the UV light irradiation, is observed only in the sample decarbonized at 1000 °C for 3h.

Figures 2 (a) and (b) shows the spectra of the Si-O-C(-H) ceramics after decarbonization at 950 and 1000 °C. During the holding at 950 °C, intensity of the PL at long wavelength region (500-600 nm) gradually increases. The peak position at 450 nm, however, does not show apparent shift even after the 3h holding. During the holding at 1000 °C, growth of PL intensity at the long wavelength region is rapid, and the peak position apparently shifts to 530 nm after the 3h holding.

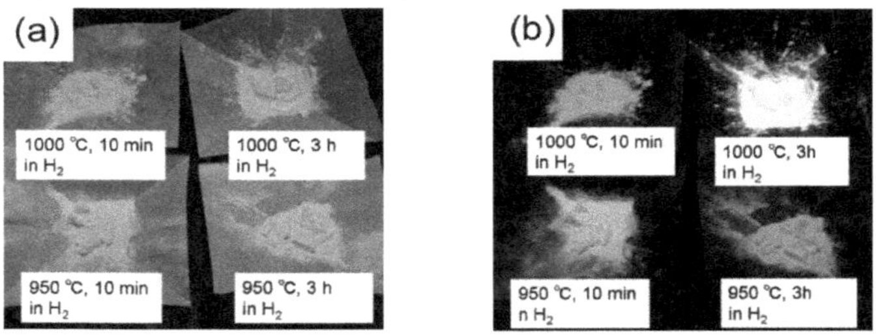

Figure 1. Photo images of Si-O-C(-H) ceramics synthesized by hydrogen decarbonization (2 μm precursor); (a) Visible light irradiation, (b) UV light irradiation.

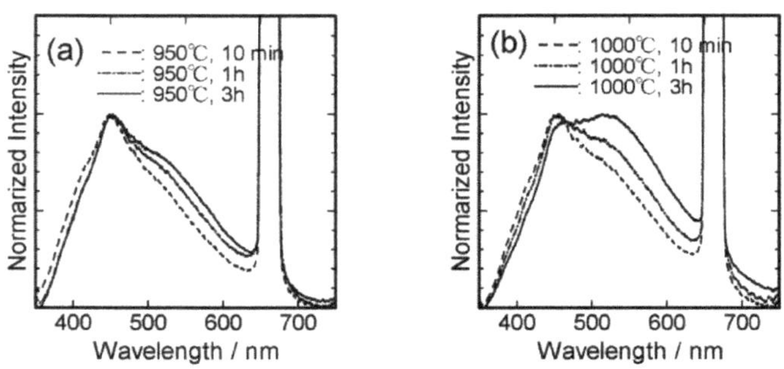

Figure 2. PL spectra of Si-O-C(-H) ceramics synthesized by hydrogen decarbonization (330 nm excitation, 2 μm precursor); (a) holding at 950 °C, (b) holding at 1000 °C.

Figures 3 (a) and (b) show phosphorescence spectra of the same Si-O-C(-H) samples. Intensity of the apparent phosphorescence increases as the holding time is extended. The normalized phosphorescence spectra exhibit exactly the same feature, which are ind ependent of the decarbonization temperature and the holding time.

Figures 4 (a) and (b) show FT-IR spectra of the samples decarbonized at 950 or 1000 °C. Most of the observed absorption bands are assigned to vibrations of silica networks. Stretching, bending and rocking vibrations of Si-O-Si appear at 1080-1050, 800 and 450 cm^{-1}, respectively.[12] On the other hand, the absorption band at 880 cm^{-1} assigned Si-C stretching is al so observed in all the spectra, which overlaps on the band of the Si-O-Si bending. During the holding at each temperature, the Si-O-Si stretching band becomes sharp. Observed changes in FT-IR spectra are, however, minor, although PL spectra of the same samples indicate apparent shift and broadening at these temperatures.

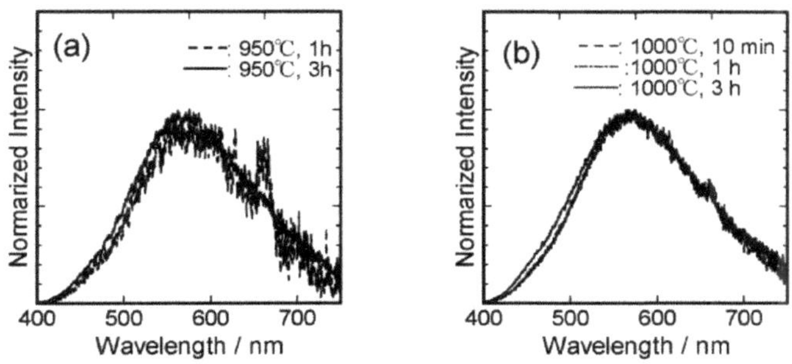

Figure 3. Phosphorescence spectra of Si-O-C(-H) ceramics (330 nm excitation, 1 ms delay time, 2 μm precursor); (a) holding at 950 °C, (b) holding at 1000 °C.

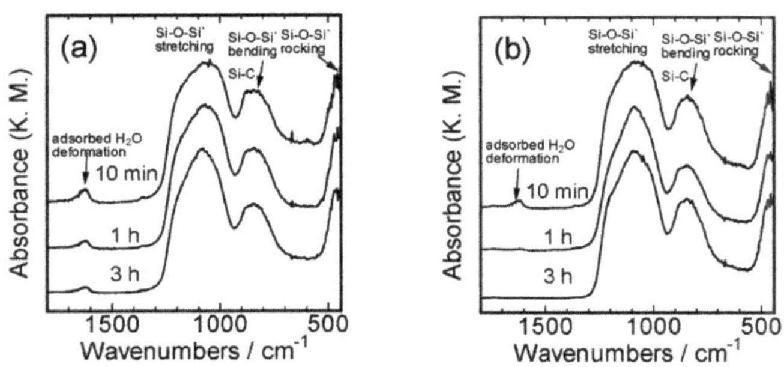

Figure 4. FT-IR spectra of Si-O-C(-H) ceramics (330 nm excitation, 1 ms shuttering time, 2 μm precursor); (a) holding at 950 °C, (b) holding at 1000 °C.

Decarbonization process using hydrogen is usually effective to precursor materials with small diameters.[13] Properties of the Si-O-C(-H) ceramics derived from the precursor microspheres with various averaged diameters are summarized in Table 1. Wh ite powders were obtained from the precursors with the averaged diameters of 2 and 6 μm. In the case of the diameter of 11 μm, however, the obtained Si-O-C(-H) ceramics become brown at 1100 °C. PL intensity also decreases and shows dark red color. Since the residual mass after the pyrolysis is a few percent large as compared with those with small diameters, the ceramics derived from the 11 μm precursor should contain excess carbon at cores of the spheres. Figure 5 (a) show PL and phosphorescence spectra of Si-O-C(-H) ceramics derived from the 11 μm spheres at 1000 °C. The peak position of the PL spectrum is at 450 nm and it shifts to 530 nm after the 3h holding. The PL spectra of the 11 μm precursor, however, possess broad tails at a "red" region (beyond 700 nm), which are not observed in the ceramics derived from the 2 μm spheres. On the other hand, the phosphorescence spectra of the 11 μm precursor show almost the same feature as those derived from the 2 μm precursor (Fig. 5 (b)).

Table 1. Properties of Si-O-C(-H) ceramics synthesized at various conditions.

Averaged diameter of precursor	Decarbonization temperature	Color under visible light (PL color)	Residual mass	Specific gravity
2 μm	800 °C	White (Blue)	83.9%	×
6 μm	800 °C	White (Blue)	81.9%	×
11 μm	800 °C	White (Blue)	83.4%	×
2 μm	1100 °C	White (Yellowish white)	78.3%	2.36
6 μm	1100 °C	White (Yellowish white)	77.9%	2.25
11 μm	1100 °C	Brown (Dark red)	80.3%	2.30

×: Measurements by a picnometer was impossible.

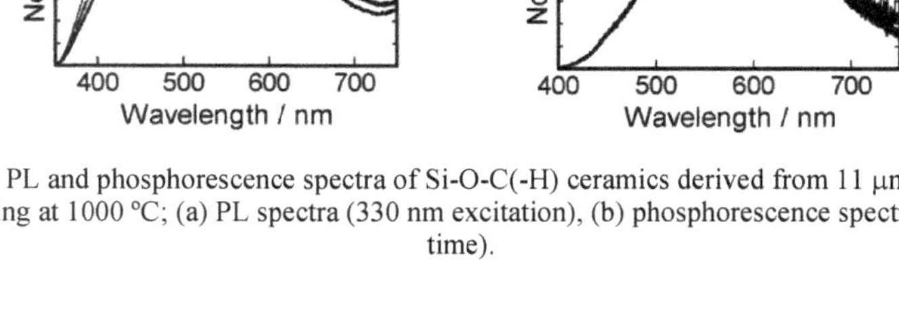

Figure 5. PL and phosphorescence spectra of Si-O-C(-H) ceramics derived from 11 μm precursor during holding at 1000 °C; (a) PL spectra (330 nm excitation), (b) phosphorescence spectra (1 ms delay time).

Figure 6 (a) shows a result of a fitting by Gaussian function on the phosphorescence spectra of the Si-O-C(-H) ceramics derived from the 2 and 11 μm precursors. The peak position of 2.18 eV and a half height line width of 0.65 eV are estimated for this long-lived PL band, which do not depend on the size of the precursor spheres. Figure 6 (b) shows a fitting result of the PL spectra of the ce ramics derived from the 11 μm precursor at 1000 °C. Most of the observed emission can be covered by the PL bands at 2.18 eV and 2.7 eV, which correspond to the long-lived PL and the short-lived PL bands previously reported.[9] As a half height line width of the short-lived PL band, 0.73 eV is derived. In addition to these main emissions, the PL spectrum of the 11 μm precursor includes the low energy "red" tail. This band below 1.8 eV is possibly assigned to PL band of excess carbon.

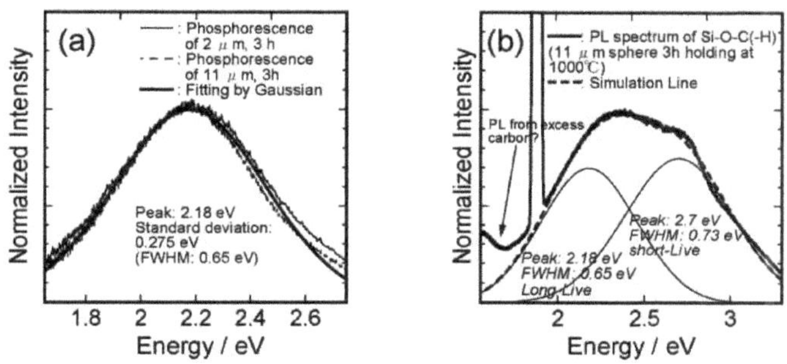

Figure 6. (a) Fitting of phosphorescence spectra by Gaussian function, (b) Fitting of the PL spectrum of the Si-O-C(-H) ceramics (11 μm precursor, 1000 °C, 3h) by assuming plural PL bands..

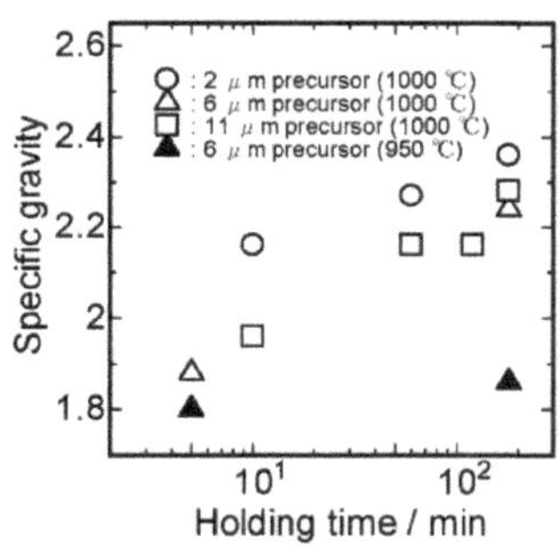

Figure 7. Specific gravities of Si-O-C(-H) ceramics synthesized at various conditions (precursor sizes, holding times, decarbonization temperatures).

Figure 7 shows specific gravity of Si-O-C(-H) ceramics derived from 2, 6 and 11 μm microspheres. At short holding times, the specific gravities are usually below 2.0. As the holding time at 1000 °C is extended, the gravity increases up to 2.23-2.36. These resulting values exceed those of amorphous silica and graphite, while they are smaller than the specific gravity (2.55) of Nicalon fiber (NL200: $SiC_{1.34}O_{0.36}$) widely commercialized.[14] Increase in the specific gravity seems to have relationship with the growth of the long-lived PL band. This long-lived PL band is known to accept quenching by molecular oxygen when BET surface area of the Si-O-C(-H) ceramics is large.[9] The relationship between the long-lived PL and the increase in specific gravity would suggest decrease of diffusion of molecular oxygen into the Si-O-C(-H) ceramics, which are densified at 950-1000 °C.

SUMMURY

Growth of the long-lived PL band in the Si-O-C(-H) ceramics is observed during the holding in a hydrogen atmosphere at 950 and 1000 °C. The growth rate of the long-lived PL band is moderate at 950 °C, and rapid at 1000 °C in appearance. The spectrum feature of this long-lived PL band is universal, which is independent of the holding time, the decarbonization temperature and the size of the precursor sphere. The peak position of 2.18 eV and the half height line width of 0.65 eV were deduced from the fitting by Gaussian function. Besides this long-lived PL band, existence of the short-lived PL band (peak position: 2.7 eV, half height line width: 0.73 eV) is suggested, which is a necessary component of the observed broad emission. A PL band assigned to excess carbon (below 1.8 eV) appears when the averaged diameter of the precursor spheres is large. The growth of the long-lived PL band seems to be accompanied by densification of the Si-O-C(-H) ceramics. After the long holding time at 1000 °C, the specific gravity exceeds 2.2, which is larger than the specific gravities of amorphous silica and carbon.

ACKNOWLEDGEMENTS

This work is partly supported by the 2011 Faculty Innovation Research Project, Graduate School of Engineering, Osaka Prefecture University.

REFERENCES

[1]M. Narisawa, S. Watase, K. Matsukawa, T. Dhomaru, K. Okamura, *Bull. Chem. Soc. Japan*, **85**, 724-726 (2012).
[2]G. Das, P. Bettotti, L. Ferraioli, R. Raj, G. Mariotto, L. Pavesi, G. D. Soraru, *Vibrat. Spectrosc.*, **45**, 61-68 (2007).
[3]I. Menapace, G. Mera, R. Riedel, E. Erdem, R.-A. Eichel, A. Pauletti, G.A. Appleby, *J. Mater. Sci.*, **43**, 5790-6 (2008).
[4]L. Ferraioli, D. Ahn, A. Saha, L. Pavesi and R. Raj, *J. Am. Ceram. Soc.*, **91**, 2422–2424 (2008).
[5]M. Fujiki, Y. Kawamoto, M. Kato, Y. Fujimoto, T. Saito, S. Hososhima, and G. Kwak, *Chem. Mater.*, **21**, 2459–2466 (2009).
[6]A. Karakuscu, R. Guider, L. Pavesi, G. D. Soraru, *J. Am. Ceram. Soc.*, **92**, 2969-74 (2009).
[7]S. Gallis, V. Nikas, H. Suhag, M. Huang, A. E. Kaloyeros, *Appl. Phys. Lett.*, **97**, 081905 (2010).
[8]S. Yu, R. Tu, A. Ito, T. Goto, *Mater. Lett.*, **64**, 2151–2154 (2010).
[9]M. Narisawa, T. Kawai, S. Watase, K. Matsukawa, T. Dohmaru, K. Okamura, A. Iwase, *J. Am. Ceram. Soc.*, **95**: 3935-3940 (2012).
[10]W. H. Green, K. P. Le, J. Grey, T. T. Au, M. J. Sailor, *Science*, **276**, 1826-1828 (1997).
[11]G. Li, L. W. Burggraf, J. R. Shoemaker, D. Eastwood, A. E. Stiegman, *Appl. Phys. Lett.*, **76**, 3373-5 (2000).
[12]J. T. Fitch, G. Lucovsky, E. Kobeda, E. A. Irene, *J. Vac. Sci. Tech. B*, **7**, 153-162 (1989).
[13]M. Takeda, A. Saeki, J. Sakamoto, Y. Imai, H. Ichikawa, *J. Am. Ceram. Soc.*, **83**, 1063-1069 (2000).
[14]K. Okamura, T. Shimoo, K. Suzuya, K. Suzuki, *J. Ceram. Soc. Japan*, **114**, 445-454 (2006).

SYNTHESIS OF HIERARCHICAL POROUS SiCO MONOLITHS FROM PRECERAMIC POLYMER IMPREGNATED WITH POROUS TEMPLATES

Xuehua Yan[1,2], Jianmei Pan[2], Xiaonong Cheng[1], Chenghua Zhang[2], Guifang Xu[2]
1. Institute for Advanced Materials, Jiangsu University, Zhenjiang, Jiangsu, China.
2. School of Materials Science and Engineering, Jiangsu University, Zhenjiang, Jiangsu, China.

ABSTRACT

SiC nanowires decorated with porous SiCO ceramics were prepared using porous materials (polyurethane sponge or wood biomass) as the templates infiltrated with preceramic polymer by pyrolysis under Ar atmosphere. The template type was an important parameter affecting the microstructure and properties of the as-prepared samples. The samples obtained at 1400 °C were composed of SiC nanocrystals and non-crystalline SiCO matrix. And SiC nanowires were formed in the pores of the SiCO ceramics. Furthermore, SiCO monoliths show a hierarchical porous structure and display a bimodal pore size distribution in the meso- and macro-size scale. The specific surface area, pore volume and average pore width of the samples can be controlled by adjusting the template type. The hierarchical porous ceramics are suitable for various applications in biological scaffold, sensors and catalyst supports.

INTRODUCTION

Porous structures can be divided into three categories according to the range of pore sizes, including macropores, mesopores and micropores. Hierarchical porous ceramics can be prepared by regulating the pore size distribution such as bimodal size distribution (micro-meso, meso-macro, micro-macro), or trimodal (micro-meso-macro). Hierarchical porous ceramics have potential applications in filters, catalyst carriers, thermal insulation materials, anti-friction materials, sensors, electromagnetic shielding materials due to their excellent performances of high-temperature resistance, thermal-shock resistance, high strength, chemical stability and high specific surface area [1-7]. Compared with the traditional methods, the fabrications of the porous ceramics from polymer precursors have some advantages such as the low temperature processing and designable molecular structure [8-9]. Up to now, hierarchical porous ceramics were produced by a controlled thermal treatment and/or the addition of suitable fillers, direct foaming and etching methods [10-13]. For example, Vakifahmetoglu et al. prepared SiCO glass monoliths possessing hierarchical porosity by periodic mesoporous organosilica (PMO) particles embedded into a foamed siloxane preceramic polymer. SiOC ceramic monoliths displayed a high amount of pores ranging in size from hundred of micrometers to a few nanometers and the components possessed a specific surface area of $137 m^2/g$ [14]. Recently, porous SiCO ceramics prepared from different templates such as polyurethane (PU) sponge, woods or poly (methyl methacrylate) (PMMA) microbeads have attracted wide attention due to the low cost and controlled open porosity [8, 15-16]. Furthermore, the porous templates can provide the growth space for the nanowires [16-17]. The formation of the nanowires made the porous ceramics have hierarchical porous structures.

The polyurethane (PU) sponges exhibit a three-dimensional net structure and possess many macropores. And wood biomass, as a natural template, processes a hierarchical cellular structure containing hollow vessels, tracheids aligned with the axis of the tree trunk and rays growing in a radial direction. These templates can be used for preparing hierarchical porous SiCO ceramics. Therefore, in this work, we prepared the hierarchical porous SiCO ceramics possessing macro and mesoporosity, using PU sponge or wood biomass as the templates impregnated with silicon resin. The phase composition of the hierarchical SiCO ceramics was investigated by XRD. The chemical structure was investigated by FTIR. The morphologies of the SiCO ceramics were observed by SEM and EDS. Lastly, the hierarchical porous structures of the SiCO ceramics obtained from different templates were studied.

EXPERIMENTAL PROCEDURE

Porous SiCO ceramics prepared from PU sponge or wood biomass

The PU sponge with a size of 30 mm×30 mm×5 mm was used as the template. The silicone resin (solid content: 48~52%, Tu-4 cup viscosity at 25 °C: 18~35s, Changzhou Jianuo Organosilicone Co., Ltd., Jiangsu, China) was used as the polymeric precursors and immersed in the PU sponge. The mass ratio of the silicone resin and PU sponge was 20:1. The silicone resin was adequately absorbed in the sponge, and then was dried at 125 °C for 8 h under vacuum drying oven. The above process was repeated three times. Then the green bodies were obtained.

Chinese fir sawdust milled into a powder of about 64μm was used as the starting material for the synthesis of the porous SiCO ceramics. Then the wood biomass and silicone resin with a mass ratio of 1:1 were mixed via magnetic stirring at room temperature. Then the mixtures were dried at 75 °C for 8 h using drying oven. The green bodies with a size of 30 mm×30 mm×5 mm were molded using the hot-pressing equipment under a pressure of 40 MPa at 80 °C for 20 min.

The sintering process was carried out in a tube furnace under a constant flow of argon gas. These green bodies were heated at a rate of 5 °C/min up to 150 °C for 15 min and a lower rate of 2 °C/min up to 400 °C for 30 min, then at a rate of 5 °C/min up to 1400 °C for 3 h. The samples were cooled down to 400 °C with a rate of 3 °C/min then down to room temperature. The samples obtained from PU sponge or wood biomass were marked as $1^{\#}$ and $2^{\#}$, respectively.

Characterization

The phase compositions were measured by X-ray diffractometer (XRD, D/max 2550 V, Japan) using a nickel filtered Cu Kα radiation produced at 35 kV and 20 mA. Infrared absorption spectra were obtained using Fourier transform infrared spectrometer (FTIR, Nicolet-460, America) in the range of 3200~500 cm^{-1}. The morphological structures of the samples were observed using scanning electron microscope (SEM, JSM-7001 F, Japan). The elemental compositions of the samples were analyzed by energy-dispersive X-ray spectroscopy (EDS) attached to SEM. The specific surface areas and pore size distributions of the samples were obtained from nitrogen adsorption-desorption measurements at 77 K using an automatic instrument (NOVA2000, USA). The specific surface areas were calculated from the isotherms by the multipoint Brunauer-Emmett-Teller (BET) method. The pore size distributions and pore volumes were determined using the Barret-Joyner-Halenda (BJH) method.

RESULTS AND DISCUSSION

XRD analysis

Fig. 1 shows the XRD patterns of the as-prepared samples obtained from different templates. XRD pattern of 1# sample obtained from PU sponge displays the characteristic of an amorphous material. The broad diffraction peak at $2\theta \approx 22°$ of the sample corresponds to the amorphous SiCO glasses. Meanwhile, the crystalline peaks at 2θ of 35.6°, 60.0° and 71.8° corresponding to the diffraction from the (111), (220) and (311) lattice planes of β-SiC (JCPDS Card No. 29-1129) were observed. It can be concluded that the SiCO ceramics prepared from PU sponge were composed of SiCO glasses and SiC crystals. However, the amorphous SiCO glasses cannot be found in 2# sample obtained from wood biomass. The diffraction peaks at 2θ of 25° and 28° corresponding to the peaks of the cristobalite (JCPDS Card No. 65-0466) appear. XRD pattern of 2# sample also shows the crystalline peaks at 2θ of 35.6°, 60.0°, 71.8° and 75.5° of β-SiC, respectively. Compared with 1# sample, the peak intensities of SiC in 2# sample is higher, indicating that the content or crystallinity of SiC phase is higher.

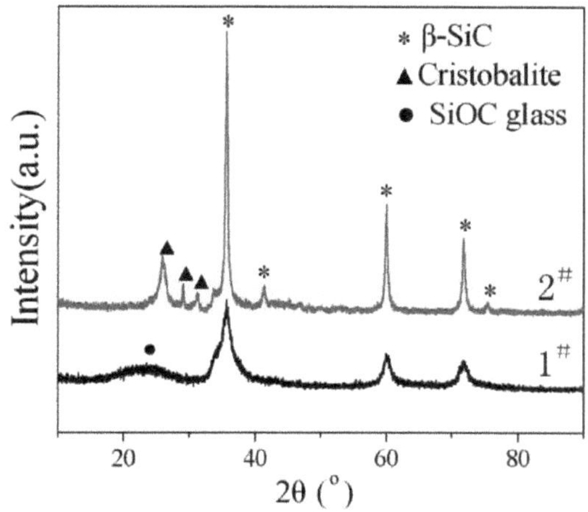

Figure 1 XRD patterns of the as-prepared samples obtained from different templates

FTIR ananlysis

Fig. 2 shows the FTIR spectra of the samples prepared from different templates. It can be found that the spectra of the samples display obvious peak in the range of 690~890 cm^{-1}. The spectrum of 1# sample shows the typical vibration band at 843 cm^{-1} assigned to the Si-C absorptions. And 2# sample presents a weak band centered at 815 cm^{-1} due to the Si-C vibration. For 2# sample prepared from wood biomass, the very weak peak at 1060 cm^{-1} is attributed to the typical Si-O absorptions. These above results indicate that the ceramic conversion of the samples obtained from different templates has been completed.

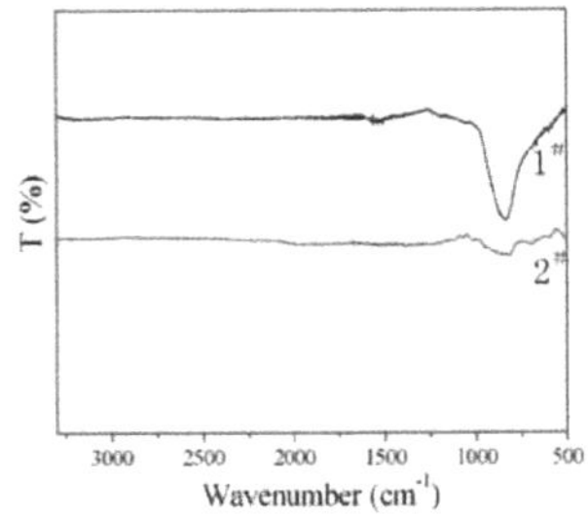

Figure 2 FTIR spectra of the as-prepared samples prepared from different templates

SEM ananlysis

The SiC nanowires with 80-150 nm diameters can be found in the pores of the SiCO ceramics (1# sample) and have different features such as curved or straight structures [16]. Fig.3 shows the SEM images of 2# sample prepared from wood biomass. The macropores-size of 2# sample are small, as shown in Fig. 3a. The low-magnification image (Fig. 3b) of 2# sample indicates the formation of the nanowires and irregular macropores with several micrometers. Fig. 3c shows the SEM image of the nanowires, indicating that the nanowires display straight or bamboo-like. The corresponding EDS spectrum of the nanowire is shown in Fig. 3d. It can be found that the nanowire is SiC nanowire. Therefore, it can be concluded that SiC nanowires can be synthesized in the pores of SiCO ceramics from different templates. Fig. 3c shows that a SiC nanowire has a round tip, whihc is typical of the vapor-liquid-solid (VLS) growth process for nanowires. SiO and CO gases derived from the pyrolysis of the silicon resion and wood biomass reacted and deposited on the pore surface to form nanowires. According to the EDS spectrum of the round tip of SiC nanowire and element analysis of wood biomass, there are iron elements, which act as catalysts.

Figure 3 SEM images of the fracture surface of the sample prepared from wood biomass (a-b), and SEM image of the nanowires (c), and the corresponding EDS spectrum of the nanowire (d)

N_2 adsorption isotherms and pore size distributions

The N_2 adsorption-desorption isotherms and the corresponding pore size distributions of the samples obtained from different templates, as shown in Fig. 4a and Fig. 4b, respectively. According to the IUPAC nomenclature, the isotherm of $1^{\#}$ sample can be classified as Type IV with H2 hysteresis. The hysteresis loops caused by the capillary condensation confirm the presence of mesopores from the accumulation of the uniform particles. The isotherm of $2^{\#}$ sample can be classified as Type IV with H3 hysteresis attributed to the formation of slit pores derived from the accumulation of the platelet-shaped particles. Furthermore, the adsorption volume of $2^{\#}$ sample is higher that of $1^{\#}$ sample. Fig. 4b shows the pore size distribution curves obtained from BJH model. It can be seen that the PSDs of the samples prepared different templates are bimodal distributions. The peaks of $1^{\#}$ sample center around 3-4 nm and 4-11 nm. $2^{\#}$ sample presents a broad pore size distribution in the range of 3-4.5 nm and 4.5-60 nm. Table 1 shows the corresponding pore features of the samples obtained from PU sponge or wood biomass. The specific surface area is 45 m^2/g and 187.5 m^2/g, respectively. And the pore volume increases from 0.069 cm^3/g to 0.293 cm^3/g. The sample prepared from wood biomass processes a higher specific surface area, mainly due to the formation of amounts of nanowires, and the existence of the natural pores of wood biomass, such as nm-scale pores of ray parenchyma cells and pits. It can be concluded that the pore features of the samples are dependent on the type of the template. These above results reflect the possible applications of the hierarchical porous ceramics in adsorption, catalysis and energy storage.

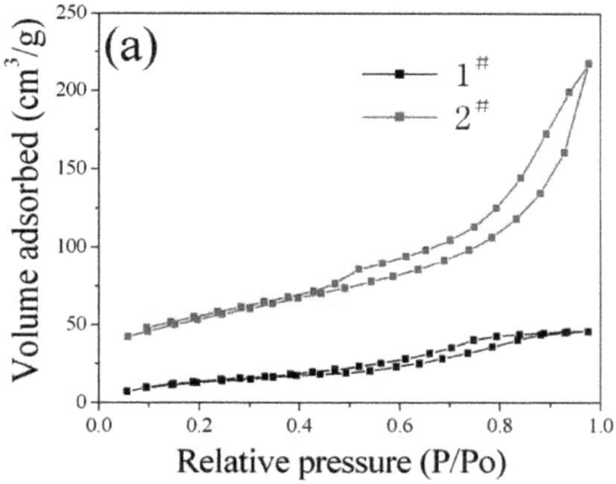

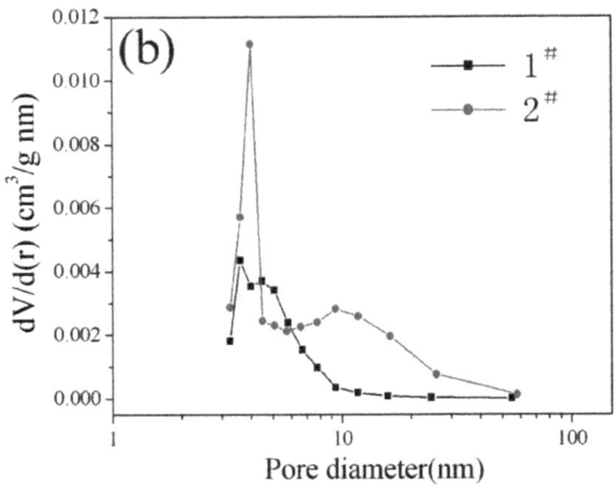

Figure 4 Adsorption-desorption isotherms of N_2 of the samples obtained from different templates (a) and corresponding pore size distribution curves calculated using the BJH method (b)

Table 1 Pore Characteristic Parameters of the Samples Obtained from Different Templates

Sample	BET surface area (m^2/g)	Pore volume (cm^3/g)	Average pore diameter (nm)
1#	45	0.069	4.09
2#	187.5	0.293	4.01

CONCLUSIONS

Hierarchical porous SiCO ceramics were synthesized from PU sponge or wood biomass as the templates, using the silicone resin as a preceramic polymer. SiC nanowires can be formed in the pores of SiCO ceramics prepared from different templates. The hierarchical porous SiCO ceramics from PU sponge have the macropores with hundreds of micrometers and mesopores in the range of 3-4 nm and 4-11 nm. The hierarchical porous SiCO ceramics prepared from wood biomass have the macropores with several micrometers and mesopores in the range of 3-4.5 nm and 4.5-60 nm. Furthermore, the sample prepared from wood biomass processes a higher specific surface area. The pore features of the samples are dependent on the type of the template. The hierarchical porous ceramics have the possible applications in biological scaffold, sensors and catalyst supports.

ACKNOWLEDGMENTS

This work is supported by the Materials Tribology Key Laboratory of Jiangsu Opening Foundation (kjsmcx07005), the Natural Science Foundation of the Jiangsu High Education (08KJD430010), the Postgraduate Research and Innovation Project of Ordinary University in Jiangsu Province (CXZZ11_0558), the Talent Foundation of Jiangsu University (09jdg033), Changshu Research Project (CG201005) and the Six Personnel Peak Project in Jiangsu Province (2011-ZBZZ045) for financial support.

REFERENCES

[1]L. Z. Fan, Y. S. Hu, J. Maier, P. Adelhelm, B. Smarsly and M. Antonietti, High Electroactivity of Polyaniline in Supercapacitors by Using a Hierarchically Porous Carbon Monolith as a Support, *Adv. Funct. Mater.*, **17**, 3083-7 (2007).

[2]Y. Gogotsi, A. Nikitin, H. Ye, W. Zhou, J. E. Fischer, B. Yi, H. C. Foley and M. W. Barsoum. Nanoporous Carbide-derived Carbon with Tunable Pore Size. *Nat. Mater.*,**2**, 591-4 (2003).

[3]J. Chmiola, C. Largeot, P. L. Taberna, P. Simon, Y. Gogotsi. Monolithic Carbide-derived Carbon Films for Micro-supercapacitors, *Science*, **328**, 480-3 (2010).

[4]Z. Liu, T. Fan, D. Zhang, X. Gong and J. Xu, Hierarchically Porous ZnO with High Sensitivity and Selectivity to H_2S Derived from Biotemplates, *Sensor. Actuat. B-Chem.*, **136**, 499-509 (2009).

[5]M. Oschatz, E. Kockrick, M. Rose, L. Borchardt, N. Klein, I. Senkovska, T. Freudenberg, Y. Korenblit, G. Yushin and Kaskel S, A Cubic Ordered, Mesoporous Carbide-derived Carbon for Gas and Energy Storage Applications, *Carbon*, **48**, 3987-92 (2010).

[6]M. Rose, Y. Korenblit, E. Kockrick, L. Borchardt, M. Oschatz, S. Kaskel and G. Yushin, Hierarchical Micro- and Mesoporous Carbide-derived Carbon as a High-performance Electrode Material in Supercapacitors, *Small*, **7**, 1108-1117 (2011).

[7]A. Zampieri, P. Colombo, G. T. P. Mabande, T. Selvam, W. Schwieger, and F. Scheffler, Zeolite Coatings on Microcellular Ceramic Foams: A Novel Route to Microreactor and Microseparator Devices, *Adv. Mater.*, **16**, 819-23 (2004).

[8]J. H. Eom and Y.W. Kim, Processing and Properties of Polysiloxane-derived Porous Silicone Carbide Ceramics using Hollow Microspheres as Templates, *J. Eur. Ceram. Soc.*, **28**, 1029-35 (2008).

[9]P. Colombo, E. Bernardo and L. Biasetto, Novel Microcellular Ceramics from a Silicone Resin, *J. Am. Ceram. Soc.*, **87**, 152-4 (2004).

[10]J. Pan, X. Cheng, X. Yan and C. Zhang, Preparation and Hierarchical Porous Structure of Biomorphic Woodceramics from Sugarcane Bagasse. *J. Eur. Ceram. Soc.*, **33**, 575-81 (2013).

[11]S. Costacurta, L. Biasetto, E. Pippel, J. Woltersdorf, and P. Colombo, Hierarchical Porosity Components by Infiltration of a Ceramic Foam, *J. Am. Ceram. Soc.*, **90**, 2172-7 (2007).

[12]C. Vakifahmetoglu, V. Presser, S. H. Yeon, P. Colombo and Y. Gogotsi, Enhanced Hydrogen and Methane Gas Storage of Silicon Oxycarbide Derived Carbon, *Micropor. Mesopor. Mater.*, **144**, 105-12 (2011).

[13]S. H. Yeon, P. Reddington, Y. Gogotsi, J. E. Fischer, C. Vakifahmetoglu and P. Colombo, Carbide-derived-Carbons with Hierarchical Porosity from a Preceramic Polymer, *Carbon*, **48**, 201-10 (2010).

[14]C. Vakifahmetoglu, P. Colombo, A. Pauletti, C. F. Martin and F. Babonneau, SiOC Ceramic Monoliths with Hierarchical Porosity. *Int. J Appl. Ceram. Technol.*, **7**, 528-535 (2010).

[15]C. Zollfrank, R. Kladny, H. Sieber and P. Greil, Biomorphous SiOC/C-ceramic Composites from Chemically Modified Wood Templates, *J. Eur. Ceram. Soc.*, **24**, 479-87 (2004).

[16]J. Pan, X. Yan, X. Cheng, Q. Lu, M. Wang and C. Zhang. Preparation of SiC Nanowires-filled Cellular SiCO Ceramics from Polymeric Precursor, *Ceram. Inter.*, **38**, 6823-9 (2012).

[17]C. Vakifahmetoglu, P. Colombo, S. M. Carturan, E. Pippel and J. Woltersdorf, Growth of One-dimensional Nanostructures in Porous Polymer-derived Ceramics by Catalyst-assisted Pyrolysis. Part II: Cobalt Catalyst. *J. Am. Ceram. Soc.*, **93**, 3709-19 (2010).

Advanced Powder Processing and Manufacturing Technologies

SOLID REACTION MECHANISM OF Li_2CO_3 AND $FePO_4$/C POWDER

Takashi HASHIZUME, Atsushi SAIKI and Kiyoshi TERAYAMA
Graduate School of Science and Engineering for Research, University of Toyama, 3190 Gofuku, Toyama, Japan

ABSTRACT

$LiFePO_4$ is used as a cathode material for rechargeable lithium-ion cell (LiFe battery). $LiFePO_4$ powder was prepared by solid phase reaction. The apparent activation energy was calculated to 152 kJ/mol by TG Ozawa method. The unreacted peak of Li_2CO_3, Fe_2O_3, Fe_3O_4 and $Li_3Fe_2(PO_4)_3$ were confirmed in the sample synthesized at 873 K to 1023 K for 8 h by using XRD. The peaks of $LiFePO_4$ were observed at 1073 K for 8 h. It's considered that diffusion controlled reaction of Li-ion was dominantly developed at a meddle field in this reaction.

INTRODUCTION

$LiFePO_4$ is used as a cathode material for rechargeable lithium-ion cell (Li-Fe) battery[1-4]. In recent years, its battery is expected as the small battery for cellular phones, mobile music players, laptop computer, etc. and large capacity battery for electric vehicles of development. In addition, Li-Fe battery has advantages such as safety, stability, low cost compared with rechargeable Li-Co, Li-Ni, their using polymer cell. Such iron as a main raw material is used; it is a great deal of resources and the inexpensive[5, 6]. There are reports that particle size, shape and crystallinity of a $LiFePO_4$ powder affect the performance of the battery[8, 9]. However, there are few reports about thermal solid phase reaction mechanism of Li_2CO_3-$FePO_4$-C system. The purpose of this study is to prepare the $LiFePO_4$ powder from the fibrous $FePO_4$, Li_2CO_3, and C mixture powder by solid phase reaction.

EXPERIMENT

Preparation of $FePO_4$

In this study, 2.875 g $NH_4H_2PO_4$ (Kanto Chemical, 99.0%) and 6.56 g $FeCl_3 \cdot 6H_2O$ (Kanto Chemical, Purity: 99.0 %) were individually dissolved in each 50 ml ion-exchanged water. The two solutions were added into a precipitate was heated at 873 K for 6 h and cooled in air to synthesize $FePO_4$ powder.

Measurements of the reaction of mixture powder by TG-DTA

Prepared $FePO_4$, commercially available Li_2CO_3 (Kanto Chemical, Purity: 99.0%) and C (Hayashi Pure Chemical, high purity reagent) powders were mixed in a mortar with a pestle (Li_2CO_3-$FePO_4$-C powder). The mixed powder was measured by thermo-gravimetry (TG-DTA, 2000s, Bulker AXS). The temperature was raised to 1173 K in an atmosphere of argon gas flow, 50 ml/min.

$LiFePO_4$ synthesized from Li_2CO_3-$FePO_4$-C mixture powder

The weight loss was observed around 723 to 973 K by TG-DTA curve of figure 1. Heating reaction temperature range of the mixture powder was applied between 723 K and 1073 K for 8 h. The mixture powder was heated under argon gas flow of 50 ml/min. The reacted powders were cooled to obtain the powder of intermediate and final product.

The crystallization of the prepared samples were investigated by using an X-ray diffract meter (XRD, LabX, Shimadzu) with 40 kV, 30 mA, Cu*Ka* radiation. The grain size and shape of the samples were observed by low vacuum scanning electron microscope (SEM, TM-1000, HITACHI).

RESULTS AND DISCUSSION

Figure 2 shows X-ray diffraction patterns of the Li_2CO_3-$FePO_4$-C powder non-reacted and reacted between 723 K and 1073 K for 8 h. There are in good agreement with the ICDD (No. 77-0094 and No. 78-1465). A diffraction peak of $Li_3Fe_2(PO_4)_3$ was observed at 873 K. The solid reaction of Li_2CO_3-$FePO_4$-C was occurred fully at 1073 K for 8h. The strong and narrow peaks show well-crystallized material by X-ray diffraction. Intermediate phase peaks of Li_2CO_3, Fe_2O_3, Fe_3O_4 and $Li_3Fe_2(PO_4)_3$ were confirmed (ICDD No. 87-1166, 87-2334 and 78-1465).The diffraction peaks of $LiFePO_4$ were observed at 923 K and more. The peak of FeO was observed at 973 K and 1023 K. Fe_2O_3 might be reduced to FeO with C. The peak of $2\theta=27.4°$ ($Li_3Fe_2(PO_4)_3$) decreased at 1023 K, however, this peak was not annihilated at 1073 K. Table 1 shows the relation of reaction temperature and lattice constant for $LiFePO_4$. The lattice parameter was only closely to ICDD at 1073 K. Therefore, it's considered that mixture phase of $Li_3Fe_2(PO_4)_3$ and $LiFePO_4$ were obtained at 1073 K for 8 h. These peaks indicated that the sample hard to be reacted entirely under 1073 K for 8 h. The more reaction temperature raise, the more unreacted peaks decrease. As a result of the XRD and TG-DTA, Li_2CO_3 was thermally decomposed, and CO_2 gas was evolved at 873 K or over. A dynamics of Li_2CO_3 decomposing to Li_2O was slowly. The evolved CO and CO_2 gases are needed to remove to promote the decomposing Li_2CO_3. The amount of CO_2 evolved was consistent to that contained in the Li_2CO_3 initial powder. The reaction formula is shown below.

$$3Li_2CO_3 + 6FePO_4 \rightarrow 2Li_3Fe_2(PO_4)_3 + Fe_2O_3 + 3CO_2 \qquad (1)$$
$$2Li_3Fe_2(PO_4)_3 + Fe_2O_3 + 3C \rightarrow 6LiFePO_4 + 3CO \qquad (2)$$

We have investigated the kinetic analysis by four TG curves using heating rate of 2, 5, 10, 15 K/min respectively about second stage in figure 1. Apparent activation energy was attempted to calculate by TG Ozawa method;

$$\text{const.} = -\log B - 0.4567\, E / RT \qquad \cdots\cdots (3)$$

E is the apparent activation energy of reaction, R is gas constant, B is heating rate of TG-DTA, *and* T is the temperature. The least-squares method was applied to each in α. The formula is shown in (4)～(7).

$$\alpha = 0.3;\quad \log B = -10.395 \times 1000 / T + 11.432 \qquad (4)$$
$$\alpha = 0.5;\quad \log B = -8.883 \times 1000 / T + 11.432 \qquad (5)$$
$$\alpha = 0.7;\quad \log B = -7.462 \times 1000 / T + 9.418 \qquad (6)$$
$$\alpha = 0.9;\quad \log B = -6.816 \times 1000 / T + 8.321 \qquad (7)$$

The results calculated equation (4) to (7) are shown in figure 3. Reaction rate on the weight loss during the reaction was shown as α. The logarithm of heating rate was plotted as a function of the reciprocal temperature. We have calculated the apparent activation energy of each reaction rate using a gradient of these straight lines. The average apparent activation energy was 152 kJ/mol. It is considered to involve that the initial stage is interfacial controlled reaction, second and final stage is diffusion controlled reaction in the TG curve. Figure 4 shows thermal kinetic calculation results of the second state in figure 1 by rapid heating. Figure 4(a) was calculated using interfacial controlled reaction and 4(b) as diffusion controlled reaction. The plots of figure 4(a) and (b) is no liner between α=0~0.4, the plots become linearly approximated at α=0.5~0.8 in figure 4(b). Both reactions were difficult to distinguish in this calculation clearly. It is considered that this calculation large activation energy was resulted in occurred parallel actions of diffusion and interfacial controlled reaction. Li_2O decomposed from Li_2CO_3 was reacted with $FePO_4$ to $Li_3Fe_2(PO_4)_3$ and Fe_2O_3 powders, and Fe_2O_3 was reduced to FeO with C. $Li_3Fe_2(PO_4)_3$, FeO and C were reacted to $LiFePO_4$. Synthesized mechanism of $LiFePO_4$ was considered appropriate for sequential reaction at next equations;

$$Li_2CO_3 \rightarrow Li_2O + CO_2 \qquad (8)$$

$$3Li_2O + 6FePO_4 \rightarrow 2Li_3Fe_2(PO_4)_3 + Fe_2O_3 \quad (9)$$
$$6Fe_2O_3 + C \rightarrow 4Fe_3O_4 + CO_2 \quad (10)$$
$$2Fe_3O_4 + C \rightarrow 6FeO + CO_2 \quad (11)$$
$$Li_3Fe_2(PO_4)_3 + FeO + C \rightarrow 3LiFePO_4 + CO \quad (12)$$

Figure 5 shows SEM image of Li_2CO_3-$FePO_4$-C mixture powder reacted at from 723 K to 1023 K for 8h. The fibrous particles were observed in figure 5 (a), these fibrous particles had about 1: 5~7 aspect ratio between a diameter and long in fig. 5 (a). The fibrous particles became misshapen in fig. 5 (b) and the aspect ratio was seemed smaller than fig. 5(a). Average sizes of the fiber were about 0.5 μm diameters and 2 μm long or smaller in fig. 5(b). The sub globular particles were observed in fig. 5 (c), (d), (e) and (f). Globular particles sizes were ~ 1 μm, and agglomerated for 1073 K.

CONCLUSION

$LiFePO_4$ powder was prepared by annealing the Li_2CO_3-$FePO_4$-C powder at 1073 K under argon gas flow. The apparent activation energy was calculated to 152 kJ/mol by TG Ozawa method. Although starting material, $FePO_4$ powder of the fibrous particles was used, $LiFePO_4$ was the aggregation of grain.

For earlier step of a chemical reaction between Li_2CO_3 and $FePO_4$/C, Li_2CO_3 was thermally decomposed to Li_2O, and the formed Li_2O was reacted with $FePO_4$ particles from contact skin. $LiFePO_4$ could be synthesized between the $Li_3Fe_2(PO_4)_3$ and FeO formed outside or surround of $FePO_4$ particles by diffusing Li-ion.

ACKNOWLEDGEMENT

This study was partially supported by Hijiri Yamada and Akiko Hirata affiliated with center for research and development in natural sciences in university of Toyama.

REFERENCES

[1] H. Liu, H. Yang, J. Li, *Electrochim. Acta* **55**, 1626-29 (2010).
[2] H. Chuangfeng, L. Pingan, J. Xueyan, *Procedia Eng* **15**, 2619 – 23 (2011).
[3] S.B.Lee, S.H.Cho, S.J.Cho, G.J. Park, S.H.Park, Y.S. Lee, *Electrochem. Commun.* **10**, 1219-21 (2008).
[4] S.B. Lee a, S.H. Cho a, S.J. Cho a, G.J. Park b, S.H. Park c, Y.S. Lee, *Electrochem. Commun.* **10**, 1219–21 (2008).
[5] J. Liu, J. Wang, X. Yan, X. Zhang, G. Yang, A.F. Jalbout, R. Wang, *Electrochim. Acta*,

54, 5656-59 (2009).
[6] Y. Xia, M.Yoshio, H. Noguchi, *Electrochim. Acta,* **52**, 240–45 (2006).
[7] T. Nagakane, H. Yamauchi, K. Yuki, M. Ohji, A. Sakamoto, T. Komatsu, T. Honma, *Solid State Ionics,* **206**, 78–83 (2012).
[8] X. Zhi, G. Liang, L.Wang, X. Ou, J. Zhang, J. Cui, *J. Power Sources,* **189**, 779–782 (2009).
[9] Y. Li, S. Zhaoa, C. Nan, B. Li a, *J. Alloys Compd.,* **509**, 957–960 (2011).

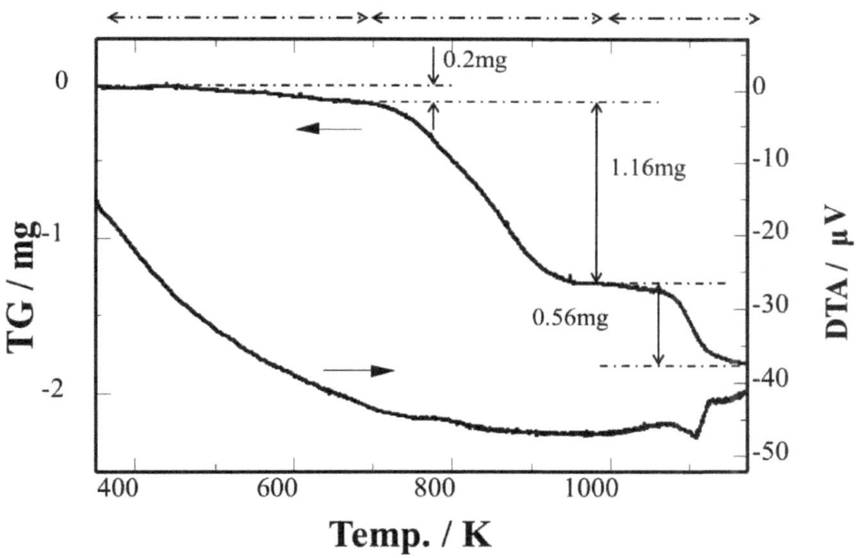

Figure 1

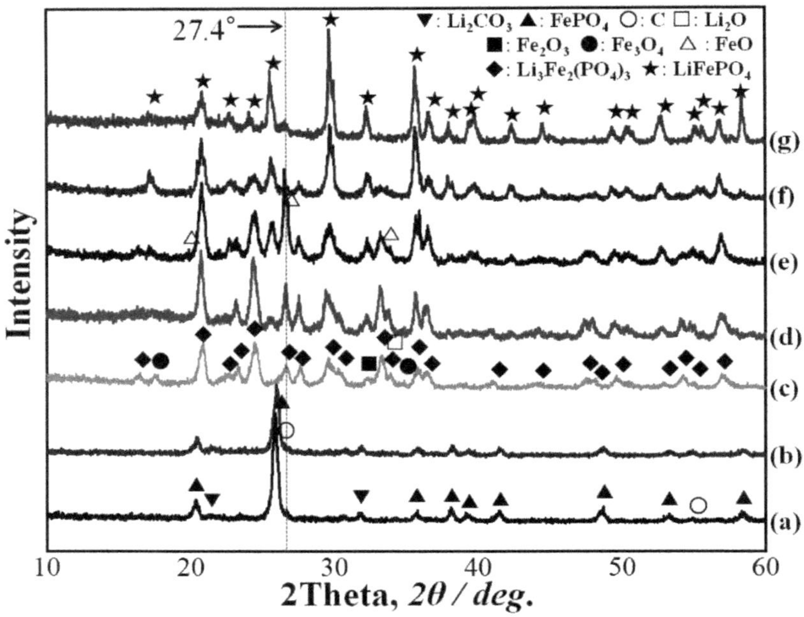

Figure 2

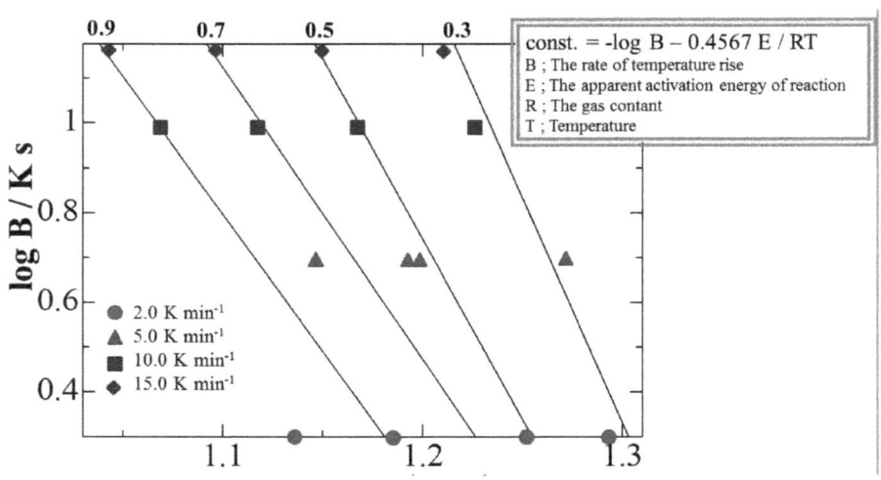

Figure 3

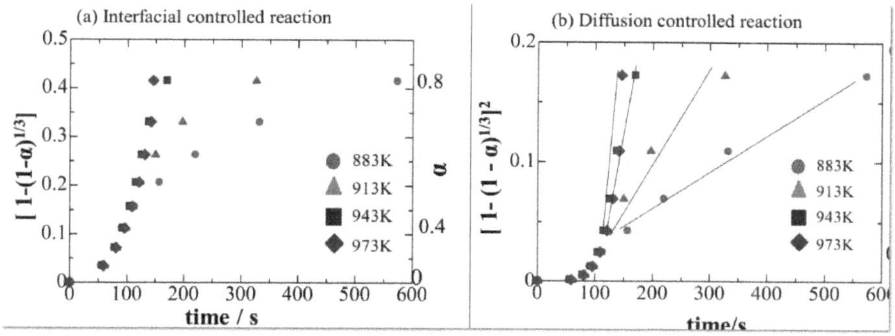

Figure 4

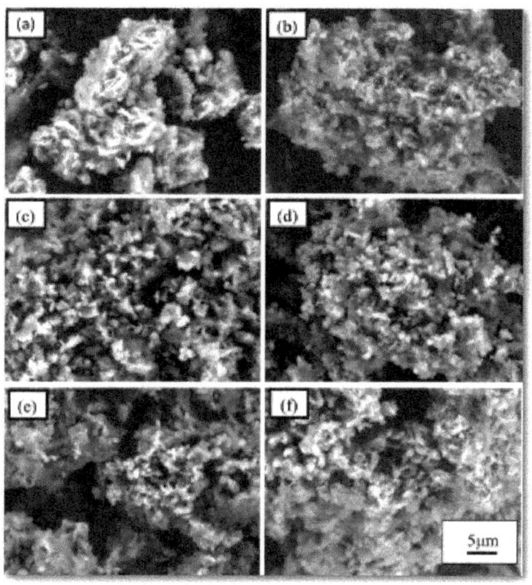

Figure 5

Table 1

	a-axis	b-axis	c-axis
ICDD of $LiFePO_4$	10.33	6.010	4.693
923 K	10.33 (4)	6.04 (1)	4.66 (3)
973 K	10.28 (1)	6.016 (5)	4.69 (1)
1023 K	10.27 (1)	6.006 (4)	4.697 (8)
1073 K	10.328 (5)	6.010 (1)	4.688 (3)

DEVELOPMENT OF NEW SYNTHESIS ROUTE OF LANTHANUM GERMANATE OXYAPATITE FROM HOMOGENEOUS AQUEOUS SOLUTION

Shouta Kitajima[1], Kiyoshi Kobayashi[2, 1], Toru Higuchi[1], Yoshio Sakka[2]

[1] Department of Applied Physics, Science University of Tokyo
6-3-1, Niijuku, Katsushika - ku, Tokyo 125-8585, Japan
[2] Materials Processing Unit, National Institute for Materials Science
1-2-1, Sengen, Tsukuba 305-0047, Japan

ABSTRACT

A new synthesis route was developed for lanthanum germanate oxyapatite by preparing homogeneous aqueous solution. The homogeneous aqueous solution was prepared from lanthanum nitrate hexahydrate, germanium dioxide with dilute ammonia, and dilute nitric acid. The precursor was obtained by addition of excessive ammonia until the pH of the aqueous solution was close to 9. X-ray diffraction analysis revealed that the lanthanum germanate oxyapatite was formed by heating at 1073 K accompanied with the impurity segregation. The oxyapatite single phase could be obtained by heating at 1373 K.

INTRODUCTION

Lanthanum germanate oxyapatite has attracted attentions to applications as new solid oxygen ion conductor for solid oxide fuel cell (SOFC) and chemical sensors[1-6]. Oxyapatites are new oxygen ion conductor family discovered by Nakayama *et al.* in 1995[7,8]. Recently, the lanthanum germanate oxyapatite has been focused on its high oxygen ion conductivity [1].

Although the lanthanum germanate oxyapatites has been synthesized by conventional solid state reaction method[1], no research exists about developing new synthesis process using aqueous solution. Considering into the low melting point (1388 K) and high volatility of GeO_2, it is worth proposing new synthesis process different from conventional solid state reaction using lanthanum oxide and germanium dioxide. We have succeeded to prepare a homogeneous aqueous solution containing both lanthanum and germanium sources. In this study, a new synthesis process for lanthanum germanate oxyapatite is developed from this homogeneous solution.

EXPERIMENTAL PROCEDURES

Preparation of the homogeneous aqueous solution and the precursor

The homogeneous aqueous solution was prepared by the following procedures (Figure 1). The raw materials are hexagonal GeO_2 (99.99 % purity, Rare Metallic Co. Ltd., Japan), lanthanum nitrate hexahydrate (99.9 % purity, Nakalai tesque Inc., Japan), ammonia solution

(Nakalai tesque Inc., Japan), nitric acid(70%, Wako pure chem. Ind. Inc., Japan), and distilled water. The lanthanum nitrate hexahydrate was dissolved in distilled water and stocked in a glass bottle. The accurate concentration of the La^{3+} in the stock solution determined by chelatometric titration[9] was about 1.5mol / l. In another beaker, stoichiometric amount of GeO_2 (about 1.43 g) was dispersed in distilled water (~200 ml). Then, the dispersed GeO_2 was dissolved in distilled water by adding the dilute ammonia solution with stirring (Figure 1). The pH value of this solution was about 9. This solution is named as Ge-aq solution hereafter.

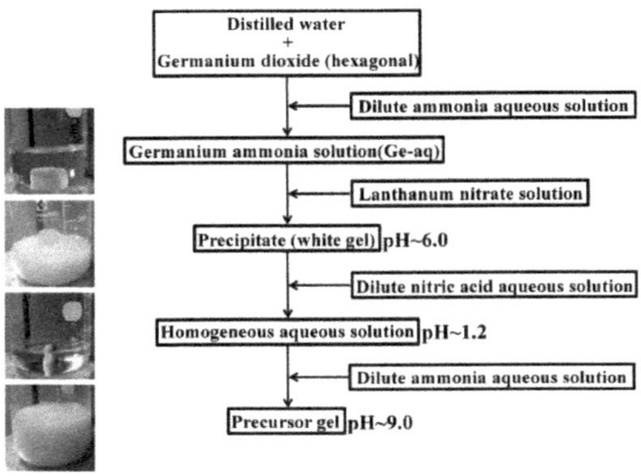

Figure 1. Flowchart of the sample preparation procedures. Left side pictures are dissolved GeO_2 by dilute ammonia (Ge-aq), white precipitation by addition of lanthanum nitrate to the Ge-aq, homogeneous solution by addition of excess nitric acid, and final precipitate by dropping of excess ammonia.

Appropriate amount of lanthanum nitrate aqueous solution was pipetted in order to fix the ratio of La and Ge at 9.67 : 6. White precipitate was formed by addition of the lanthanum nitrate aqueous solution. The pH of this condition was about 6. The white precipitate was dissolved again by dropping the dilute nitric acid solution as shown in Figure 1. The pH value of the homogeneous solution was about 1.2.

The precipitate was obtained by addition of excessive ammonia into the homogeneous aqueous solution until the pH reached about 9. The white precipitate was filtered and then washed by distilled water. The washing and filtering procedures were repeated twice to ensure the removal of adsorbed ammonia and ammonium nitrate. After the washing, the precipitate was dried at 373 K in air. The dried precipitate was pulverized into powder using an agate mortar and a pestle, and the powder was used as precursor.

Characterization

The precursor powder was heated at 873 K, 1073 K, 1273 K, and 1373 K for 3 h in air in order to confirm the formed phases. These heat treated samples are labeled as LG-873, LG-1073, LG-1273, and LG-1373, respectively. The formed phases were analyzed by powder X - ray diffraction (XRD) (RINT-2000, Rigaku Co., Japan). The powder morphology of the precursor and heated samples were observed by scanning electron microscope (SEM) (JSM-6500F, JEOL Co., Japan). Thermal analyses for the precursor powder were performed in a gas mixture of 80% N_2 and 20% O_2 at a heating rate of 10 K/min using TG–DTA instruments (TG8120, Rigaku Co., Japan).

RESULTS AND DISCUSSION

XRD patterns

The XRD patterns of the precursor, LG-873, LG-1073, LG-1273,and LG-1373 are shown in Figure 2. The XRD pattern of the precursor is composed of two kinds broad peaks: the typical halo pattern and crystalline peaks assigned to be $La(OH)_3$. The halo pattern is likely to be due to a neutralized salt formed by mixing of the lanthanum nitrate and Ge-aq solutions because of the following facts. White precipitate is formed when the lanthanum nitrate and Ge-aq solutions are mixed together. This white precipitate is dissolved by the addition of nitric acid. No precipitate is formed when excess nitric acid is added into the Ge-aq solution. Other details of the neutralized salt will be published elsewhere.

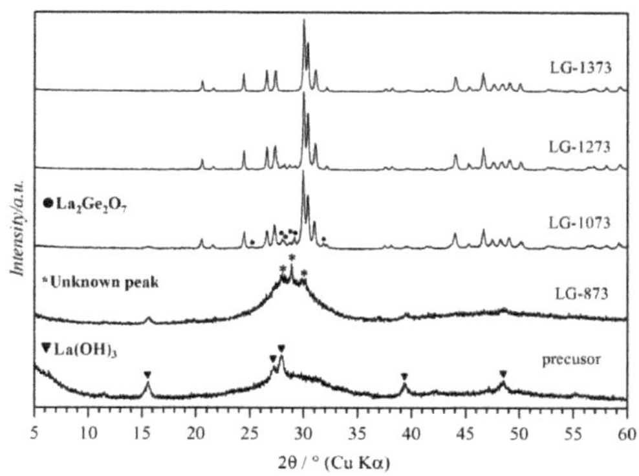

Figure 2. Powder X-ray diffraction patterns of the precursor, LG-873, LG-1073, LG-1273, and LG-1373. All of the patterns were collected at room temperature.

By the heating at 873 K, the XRD peak intensity from the $La(OH)_3$ decreases, while a broad halo peak is grown around $2\theta = 29°$ in addition to relatively sharp peaks at $2\theta =$

28°, 29°, and 30° as shown by the pattern of the LG-873 in Figure 2.

Although the XRD peaks assigned to be due to the oxyapatite phase are found in the sample heated above 1073 K, weak peaks from impurity phase are found in case of the LG-1073 and LG-1273 samples. The peak positions are consistent to those of $La_2Ge_2O_7$[10]. On the other hand, the XRD peaks of LG-1373 are composed by only the oxyapatite phase.

Considering to the fact that the precursor is a mixture of amorphous and $La(OH)_3$ phases, our synthesis process is referable to a coprecipitation method.

Morphology of the prepared powders

The SEM images of the precursor and LG-1073 powders are shown in Figure 3. The precursor powder is mainly composed of formless aggregated particles with 1μm order size. On the other hand, the LG-1073 powder is found to be composed of the nano-sized particles with about 100 nm order. Comparing the morphology of the precursor powders, lanthanum germanate oxyapatite particles seem to be formed by release of the chemical species such as water and carbon dioxide. By the chemical reactions, the powder became small even though the temperature increased.

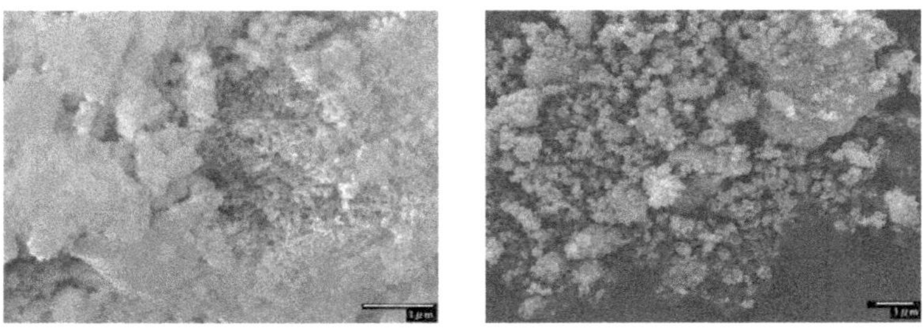

Figure 3. Morphology of the precursor and LG-1073 powders.

Thermal analysis

The TG-DTA curves of the precursor powder are shown in Figure 4. Gradual weight change appears during heating from room temperature to 1373 K accompanied with broad endothermic and exothermic peaks. Large weight loss around 473 K with broad endothermic heat is ascribed to the release of adsorbed water. Referring to the TG-DTA curves of $La(OH)_3$ [11], the endothermic peaks at 623-683 K (shown by arrow (a) in Figure 4), 773-853 K (shown by arrow (b) in Figure 4), and 943-1083 K (shown by arrow (c) in Figure 4) correspond to the decomposition reactions from $La(OH)_3$ to LaOOH, from LaOOH to La_2O_3, and from lanthanum hydroxycarbonate to La_2O_3, respectively. As similar to the decomposition of $La(OH)_3$ precipitated from the aqueous solution by addition of hexamethylenetetramine [11], small amount of CO_2 seem to be contained in our samples. The contamination of CO_2 in our samples was confirmed by the IR spectrum. Small exothermic peak appeared at 1073-1123 K is due to the formation of the lanthanum germanate oxyapatite and impurity phases by the crystallization from amorphous. However, the crystallization reaction seemed to be accompanied with small weight loss as shown in Figure 4; the crystallization reaction might proceed with releasing chemical species such as water and CO_2.

Hence, the formation mechanism of the lanthanum germanate oxyapatite from our precursor is not a crystallization from an amorphous phase.

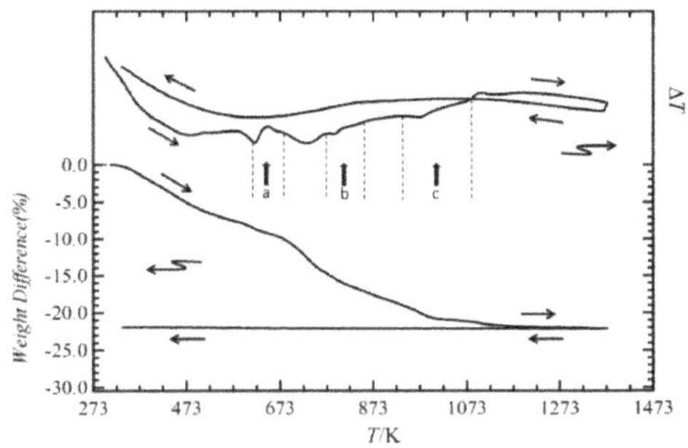

Figure 4. TG-DTA curve measured using the precursor powder.

CONCLUSION

We succeeded to prepare a homogeneous aqueous solution containing lanthanum and germanium sources. The precursor could be obtained by addition of excess ammonia to the homogeneous aqueous solution. The precursor was a mixture composed to amorphous phase and $La(OH)_3$. The lanthanum germanate oxyapatite was synthesized by heating above 1073 K accompanied with the impurity phase precipitation. Single phase of the lanthanum germanate oxyapatite could be obtained by heating at 1373 K. From the results of SEM observation and TG-DTA analysis, it was concluded that the lanthanum germanate nanopowder was formed by heating at 1073 K accompanied with weight loss by chemical reaction.

REFERENCES

[1]H. Arikawa, H. Nishiguchi, T. Ishihara, and Y. Takita , Oxide Ion Conductivity in Sr-doped $La_{10}Ge_6O_{27}$ Apatite Oxide, *Solid State Ion.* , 136-137, 31-7(2000)

[2]S. Nakayama, M. Sakamoto, Ionic Conductivities of Apatite-type $La_x(GeO_4)_6O_{1.5x-12}$ (X=8-9.33) Polycrystals, *J.Mater.Sci.Lett.* , 20,1627-29(2001)

[3]J.E.H. Sansom, L. Hildebrandt, and P.R. Slater, An Investigation of the Synthesis and Conductivities of La-Ge-O Based Systems. , *Ionics*, 8 ,155-60(2002)

[4]L.L. Reina, M.C.M. Sedeno, E.R. Losilla, A. Cabeza, M.M. Lara, S. Bruqe, F.M.B. Marques, D.V. Sheptyakov, and M.A.G. Aranda, Crystalchemistry and Oxide Ion Conductivity in the Lanthanum Oxygermanate Apatite Series, *Chem. Mater.* , 15,2099-108(2003)
[5]P.R. Slater, J.E.H. Sansom, and J.R. Tolchard, Development of Apatite-Type Oxide Ion Conductors, *The Chem Record.* , 4, 373-84(2004)
[6]E. Kendick, M.S. Islam, and P.R. Slater, Developing Apatites for Solid Oxide Fuel Cells: Insight into Structual, Transport and Doping Properties, *J. Mater. Chem.* , 17,3104-11(2007)
[7]S. Nakayama, H. Aono, and Y. Sadaoka, Ionic Conductivity of $Ln_{10}(SiO_4)_6O_3$(Ln=La, Nd, Sm, Gd and Dy), *Chem. Lett.* , 24, 431-32 (1995)
[8]S. Nakayama, T. Kageyama, H. Aono, and Y. Sadaoka, Ionic Conductivity of Lanthanoid Silicates, $Ln_{10}(SiO_4)_6O_3$ (Ln=La, Nd, Sm, Gd, Dy, Y, Ho, Er and Yb), *J. Mater. Chem.* , 5,1801-05(1995)
[9]K. Kobayashi, Y. Matsusita, N. Igawa, F. Izumi, C. Nishimura, S. Miyoshi, Y. Oyama, and S. Yamaguchi, Water-based Sol-gel Synthesis and Crystal Structure Refinement of Lanthanum Silicate Apatite, *Solid State Ion.* , 179,2209-15(2008)
[10]G. Vetter, F. Queyroux, Détermination Structurale de la Forme de Haute Température Ducomposé $La_4[Ge_3O_{10}][GeO_4]$: Analyse Comparative des Structures Tricliniques des Germanates de Terres Rares de Formule Globale $Ln_2Ge_2O_7$, *J. Solid State Chem.* , 73,287-97(1988)
[11]M. Ozawa, R. Onoe, and H. Kato, Formation and Decomposition of Some Rare Earth (RE = La, Ce, Pr) Hydroxides and Oxides by Homogeneous Precipitation, *J. Alloys Compd.* , 408–412,556–59(2006)

MAGNETIC ORIENTATION OF BISMUTH NANO-PARTICLES IN A TRANSPARENT MEDIUM

Naoyuki Kitamura[1], Kohki Takahashi[2], Iwao Mogi[2], Satoshi Awaji[2] and Kazuo Watanabe[2]
[1] National Institute of Advanced Industrial Science and Technology
Ikeda, Osaka, Japan
[2] Institute for Materials Research, Tohoku University
Sendai, Miyagi, Japan

ABSTRACT

We have prepared spherical and plate-like bismuth nano-particles by the polyol process. The sphere size was varied from 200 to 400 nm by the addition of polyvinylpyrrolidone to inhibit the aggregation of atoms. Nano-plates 800 nm wide and 50 nm thick were formed by adding a probe-level ferric chloride. The crystal axis of the nano-spheres (diameter: 350 ± 50 nm) dispersed in polysiloxane was oriented under high magnetic fields because of the anisotropic susceptibility of bismuth. The anisotropic susceptibility was estimated as 3.6×10^{-5} from a numerical simulation of the field dependence of the (110) peak in the XRD pattern. The value was slightly smaller than that for a bulk single crystal. Composites of bismuth nano-spheres and plates dispersed in poly(methyl methacrylate) were prepared under high magnetic fields. The composites showed different transmittance depending on whether the direction of the propagation of light was parallel or perpendicular to the applied magnetic field, because of the anisotropy of the dielectric constants and the shape of the nano-plates.

INTRODUCTION

The optical anisotropy and non-linearity of composites which contain dispersions of metallic or semiconducting nano-particles with sub-micrometer or smaller diameters mean they are particularly interesting materials for optical devices. For example, the absorption spectra of elongated silver nano-particles show optical anisotropy caused by the difference in plasmon resonance frequencies along the two directions of the particle [1,2]. The silver nano-particles are oriented by mechanical elongation of the composite at high temperatures, and the resulting composite is used as a polarizer. Magnetic orientation is an attractive process for constructing anisotropic structures in composite and homogenous materials and also for creating new electrical, optical, and mechanical functions related to the structure and properties of the particle. The motive force responsible for the orientation is the torque generated by the anisotropic magnetic energy of the particles in the magnetic field. When the anisotropic susceptibility of the particles is large, the crystal axis orients along the magnetic field, regardless of the shape of the particle. Advances have been made in generating high magnetic fields with superconducting magnets during the last two decades, which has allowed the magnetic orientation and alignment of crystallites, polymer chains, and fibers to be extensively studied [3-7]. Bismuth is a half-metal with an unusually large anisotropic susceptibility [8]. Therefore, it can be oriented under magnetic fields of several teslas [9]. Transparent materials containing oriented bismuth nano-particles are expected to display new optical and electrical functions because of the anisotropic electric, dielectric, and electromagnetic properties of bismuth.

In this paper, we have prepared bismuth nano-particles of various shapes by the polyol process. The dependence of the shape and size of the particles on additives was examined. X-Ray diffraction was used to observe the behavior of the magnetic orientation of the bismuth nano-particles and a numerical simulation was used to discuss the results. Moreover, we demonstrated the optical anisotropy of composites of bismuth nano-particles dispersed in poly(methyl methacrylate) (PMMA), which were prepared under a high magnetic field.

EXPERIMENTAL PROCEDURE

Preparation of bismuth nano-particles

Bismuth nano-particles were prepared by the polyol process, which was similar to that reported by Wang and Kim [10]. A flow chart of the sample preparation is shown in fig. 1. Bismuth trichloride (70 mg, Kojundo chemical, 99.99%) was dissolved in ethylene glycol (50 mL, Kishida, 99.5%) with NaOH (25-150 mg, Kishida, 95%) and polyvinylpyrrolidone (0.2-2 g, Aldrich, M_w ~55,000) in a glove box and stirred vigorously under a nitrogen flow for 3 h. Polyvinylpyrrolidone was used to inhibit the aggregation of atoms and as a template to control the shape and size of the nano-particles. The mixture was placed in a three-necked flask fitted with a reflux condenser, then heated to 190 °C over 1.5 h and stirred at 190 °C for 1 h under nitrogen. The black bismuth nano-particles were obtained after the heat treatment. The suspension was poured into ice-cooled ethylene glycol (75 mL) and then collected by centrifugation. The nano-particles were cleaned by sonication and centrifugation in ethanol. The nano-particles were characterized by scanning electron microscopy (SEM; Hitachi, S-4800). The plate-like particles were prepared by adding a probe-level amount of ferric chloride and sodium hydroxide (70 mg).

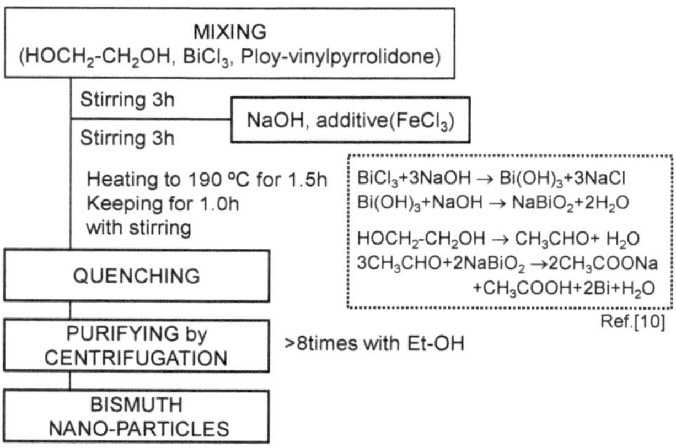

Figure 1. Flow chart of the preparation of bismuth nano-particles by the polyol process.

X-ray diffraction measurements

An XRD pattern was obtained under high magnetic fields of up to 5 T using the 5T-CSSM split-type superconducting magnet at the High Field Laboratory for Superconducting Materials, Institute for Materials Research, Tohoku University. Figure 2 shows a diagram of the in situ XRD observation system [11]. Bismuth nano-spheres (10 mg), which were 350 ± 50 nm in diameter, were dispersed in polydimethylpolysiloxane (0.75 mL, Dow Corning, SRX-310) by ultrasonication. The viscosity of the sample was estimated to be about 0.1 Pa·s under experimental conditions at 293 K. It was confirmed by optical microscopy that the particles did not form aggregates in the liquid. The sample was put in a plastic cell with a volume of 12 × 18 × ~1 mm ($h \times w \times t$). The cell had a 125-μm-thick polyimide film window for the XRD measurements. The X-ray beam (Cu-Kα: 30 kV-40 mA) had a θ-2θ arrangement. A magnetic field of up to 5 T was applied vertically.

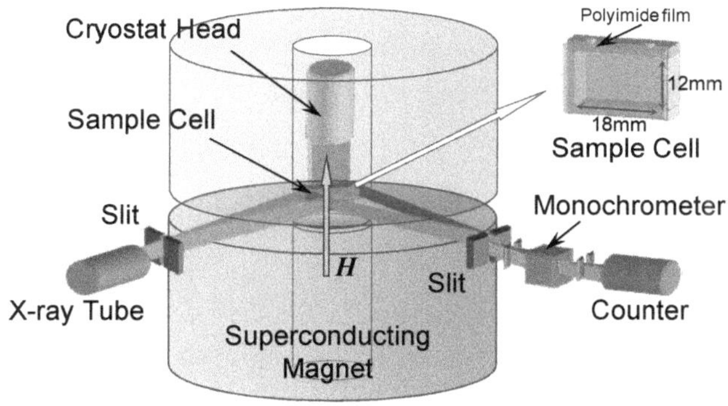

Figure 2. Schematic diagram of the split-type superconducting magnet equipped with X-ray diffraction apparatus. Inset shows detail of the sample holder, consisting of a thin plastic cell with a polyimide film X-ray window.

Preparation and optical properties of bismuth nano-particle dispersions in PMMA

Bismuth nano-spheres and nano-plates dispersions in PMMA were prepared in a super conducting magnet (Japan Magnet Technology, JMTD-10T100N1-PC, bore diameter = 100 mm) equipped with a thermostatic chamber. A solution of benzoyl peroxide (0.5 g) in undistilled methyl methacrylate (MMA; 100 g) monomer was heated at 90 °C for 4 h to obtain a partially polymerized viscous liquid. Bismuth nano-spheres (1.1 mg) were dispersed in a portion of the liquid (7 mL) by ultrasonication. The product was placed in a vial, sealed in a sample tube, and put in the magnet. The sample tube was kept at 50 °C under an 8 T field for about 8 h until polymerization was complete. Slices (2 mm thick) were cut from the polymerized composite parallel and perpendicular to the direction of the applied magnetic field. The samples were polished to an optical grade with a conventional procedure for optical glass. It was confirmed using an optical microscope that the nano-spheres did not form aggregates in the composite and that the average distance between particles was about 30 μm. The transmittance of the sample plates was measured with a conventional spectrophotometer (Hitachi, U-4000) from 200 to 2400 nm with and without polarizer.

RESULTS AND DISCUSSION

Size and shape of nano-particles

The size of the spherical nano-particles was varied from 200 to 400 nm by the addition of 25 to 150 mg sodium hydroxide, whereas the amount of polyvinylpyrrolidone was constant. Figure 3(a) shows a representative SEM image of the bismuth nano-spheres obtained by adding 70 mg of sodium hydroxide. The size distribution broadened when less than 40 mg of sodium hydroxide was added. Only spherical particles were formed when ferric chloride was not added. Plate-like particles formed when a probe-level amount of ferric hydrate was added. The shape of the plate-like particles varied with the amount of sodium hydroxide when more than 0.7 g of polyvinylpyrrolidone was added. Hexagonal plate-like particles 800 nm wide and 50 nm thick were obtained by adding 50-100 mg sodium hydroxide (fig. 3(b)). Plate-like particles accounted for 60-80% of the nano-particles, and the other particles were spherical. The fraction of plate-like particles decreased and that of the spherical particles increased when less than 50 mg of sodium hydroxide was added. In contrast, when more than 100 mg of sodium hydroxide was added thin rod-like particles were formed (fig. 3(c)). The maximum fraction of rod-like particles was about 40%.

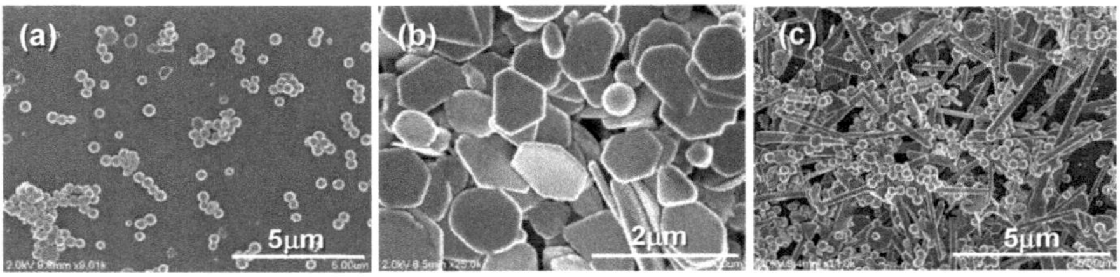

Figure 3. SEM images of (a) spherical, (b) hexagonal plate-like, and (c) thin rod-like bismuth nano-particles prepared by the polyol process.

Magnetic orientation of bismuth nano-spheres

Figure 4 shows the XRD patterns of bismuth nano-spheres 350 nm in diameter, which were measured at 0 and 5 T. All peaks observed in the XRD patterns were assigned to a rhombohedral

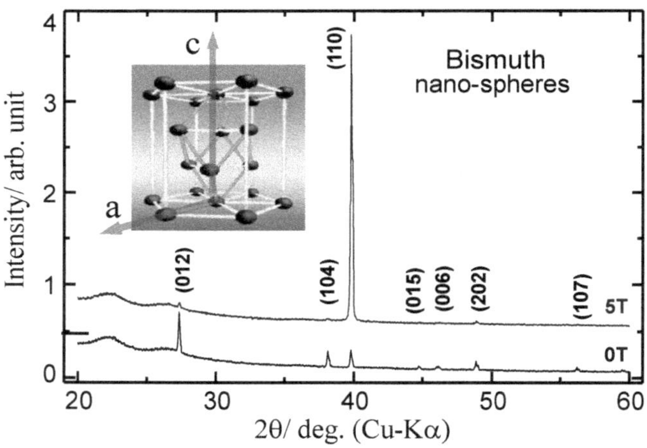

Figure 4. XRD patterns of bismuth nano-spheres 350 nm in diameter dispersed in polydimethylpolysiloxane under 0 and 5 T. Inset shows the crystal structure of bismuth for the rhombohedral phase.

bismuth phase. The broad peak at lower angles was caused by the diffraction of the sample cell and polydimethylpolysiloxane. The peak assigned to (110) increased and the peak assigned to (104) decreased at 5 T. This is consistent with the results reported by Kohama et al. [9]. The change in peak intensity indicates the c-axis was oriented in the direction of the applied field. The intensities of the two peaks are plotted against the applied magnetic field in fig. 5. The (110) and (104) peaks were almost saturated above 4 T. Magnetic orientation was achieved when the magnetic anisotropic energy, $\Delta\chi\mu_0H^2/2$, was larger than the thermal disturbance, k_BT, where $\Delta\chi$ is the anisotropic susceptibility, μ_0 is the vacuum permeability, H is the applied magnetic field, k_B is the Boltzmann constant, and T is the temperature of the system. Because metallic bismuth has a uniaxial anisotropy, the anisotropic susceptibility is described by $\Delta\chi = \chi_{//} - \chi_{\perp}$. However, the magnetic easy axis of the particle is distributed over different angles at lower magnetic fields. The distribution probability, $P(\theta,H,T)$, of the magnetic orientation is described by Boltzmann distribution by the following equations

$$P(\theta,H,T)d\theta = \frac{\exp[-E(\theta,H)/kT]d\theta}{\int_0^{\pi}\exp[-E(\theta,H)/kT]d\theta},$$

$$E(\theta,H) = -\frac{1}{2}n\mu_0 H^2[\chi_{\perp} + (\chi_{//} - \chi_{\perp})\cos\theta],$$

where $E(\theta, H)$ is the magnetic anisotropic energy of a particle where the easy axis makes an angle θ with the direction of the magnetic field, and n is the mole number of bismuth atoms in the particle. The theoretical probability was calculated for each magnetic field, and the probability at $\theta = 0$ was fitted to the (110) peak intensities by the least squares method (fig. 5). The theoretical results were consistent with the experimental data, and the estimated anisotropic susceptibility, $\Delta\chi$, was calculated as 3.6×10^{-5}, which was slightly smaller than the value for bulk bismuth, 5.3×10^{-5} [8]. Although the reason for the difference is not clear yet, it may be caused by the size effect. Furthermore, the differences may arise from impurities in the particles; Schoenberg and Uddin have reported that the susceptibility of bismuth is altered substantially by its effective valence number [8].

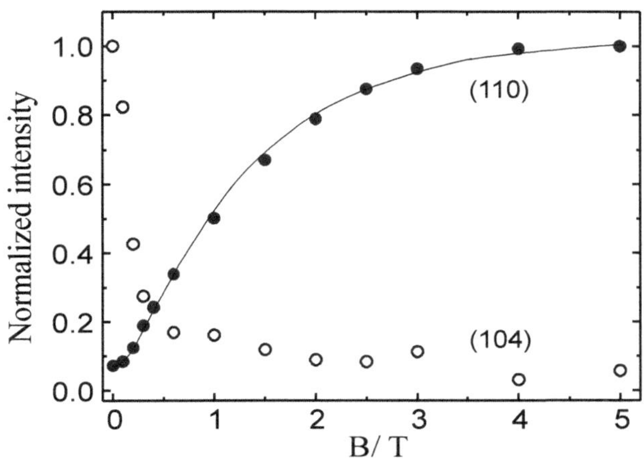

Figure 5. Plot of the (110) and (104) XRD peak intensities against the applied magnetic field. Solid line represents the results of the numerical simulation of magnetic orientation using the Boltzmann distribution.

Optical anisotropy of bismuth nano-sphere and nano-plate PMMA composites

The bismuth nano-spheres were oriented in the direction of the magnetic field above 4 T. Because the mass of the nano-plate is in the same order as that of a nano-sphere 350 nm in diameter, the behavior of the plates under a magnetic field should be the same as the spheres. Therefore, the nano-sphere and nano-plate PMMA composites polymerized at 8 T should be oriented in the direction of the applied magnetic field. Figure 6 shows the transmission spectra of the nano-sphere composite where the direction of the unpolarized light propagation was parallel or perpendicular to

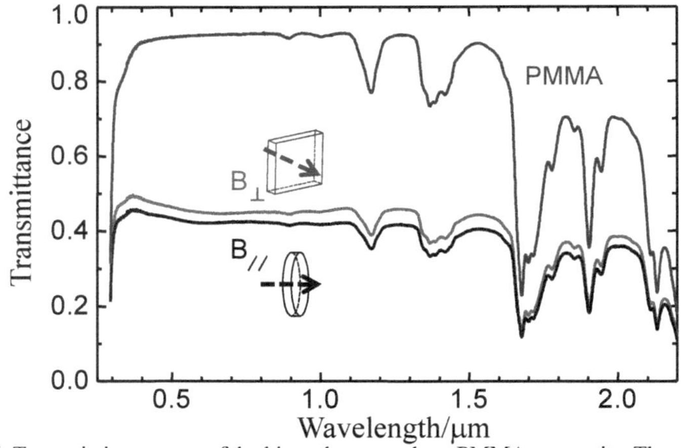

Figure 6. Transmission spectra of the bismuth nano-sphere PMMA composite. The spectrum of undoped PMMA is shown for comparison. Red and black lines represent the spectra for light propagation parallel and perpendicular to the direction of applied magnetic field, respectively.

the field direction during polymerization. The transmittance of the composite decreased over the entire region compared with undoped PMMA. Moreover, the transmittance for the light propagation perpendicular to the field direction was higher than that parallel to the field, although the particles were geometrically isotropic. The decrease in transmittance over the entire region originated from the light scattering of the particles in PMMA [12]. The scattered light is not detected by conventional spectrophotometers. The origin of the difference in transmittance between two directions is probably the anisotropy of the optical constants of bismuth crystals. The optical constants of bismuth, n and k, have been reported by Atkinson [13] and Lenham [14]. The absorption coefficient, k, parallel to the c-axis is 10-20% larger than that perpendicular to the axis. Therefore, the difference in transmittance is qualitatively consistent with the anisotropy of the optical constant.

Figure 7 shows the transmission spectra of the nano-plate composite, where the direction of the propagation of unpolarized light was parallel or perpendicular to the field direction during polymerization. The transmittance for the light perpendicular to the field direction was higher than that for light parallel to the field direction. The difference in transmittance between the two directions was larger than that for the nano-sphere composite. The anisotropy of the optical constants is also probably caused by the anisotropy of the optical constants of bismuth crystals, as with the spherical particles. Furthermore, a large difference in between the cross sections parallel and normal to the plate surface increased the difference in transmittance. Anomalous behavior was also observed in the transmittance of polarized light parallel to the plate surface. In fig. 7, the dashed lines at longer wavelengths for the sample with a horizontally oriented nano-plate surface show that the transmittance for horizontally polarized light (HH) was smaller than that for vertically polarized light (HV). The width of the plate was in the same order as the wavelength and the thickness was smaller. Therefore, the interaction of the polarized light with the plate in the HV arrangement should be smaller than that for the HH arrangement, as in the case of a wire-grid polarizer. This is consistent with our experimental results. However, the magnitude of the transmittance for the two polarized lights swapped over at 0.7 μm. This may be caused by the overlap of a broad absorption band with a peak at 550 nm in the spectrum for vertically polarized light, although the origin of the absorption band is not clear.

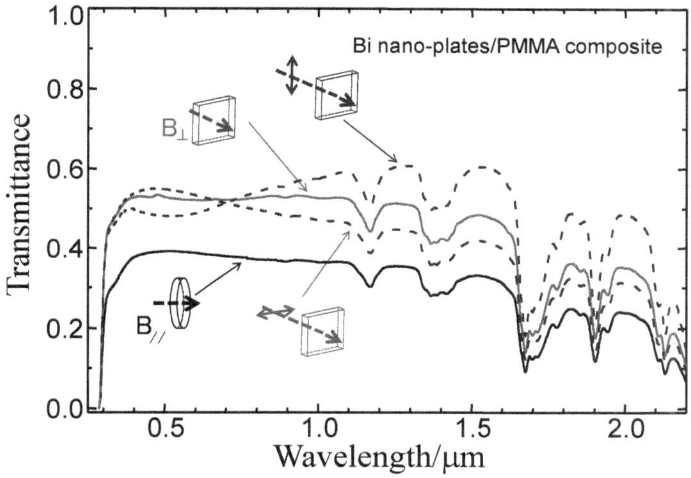

Figure 7. Transmission spectra of the bismuth nano-plate PMMA composite. Red and black lines represent the spectra for the propagation light parallel and perpendicular to the direction of applied magnetic field, respectively. Dashed lines indicate spectra for vertically and horizontally polarized light for the sample where a vertical magnetic field was applied during polymerization.

CONCLUSION

We have prepared spherical, plate-like, and thin rod-like bismuth nano-particles by the polyol process. The anisotropic susceptibility of the spherical particles was determined from their orientation behavior under high magnetic fields. We dispersed oriented spherical and hexagonal plate-like bismuth particles in transparent PMMA under a magnetic field of 8 T. The composite showed different transmittances depending on whether the direction of light propagation was parallel or perpendicular to the applied field, indicating the optical constants of the nano-particles were anisotropic. The difference for the plate-like particle composite was larger than that for the spherical particle composite. This was caused by the large difference in cross sections between the directions normal and parallel to the plate surface.

ACKNOWLEDGEMENTS

This work was supported by Grant-in-Aid for Scientific Research (C) (24560832) from the Ministry of Education, Culture, Sports, Science and Technology of Japan. The authors are deeply grateful to Dr. K. Kintaka and Dr. K. Fukumi for helpful discussions.

REFERENCES

[1]Stookey, S. D. and Araujo, R. J., Selective Polarization of Light Due to Absorption by Small Elongated Particles in Glass, *Appl. Opt.* **7**, 777-780 (1968).

[2]Sawada, T., Ando, S. and Sasaki, S., Optical Anisotropy of Uniaxially Drawn and Silver-dispersed Polyimide Films, *Appl. Phys. Lett.* **74**, 938-940 (1999).

[3]Torbet, J., Freyssinet, J.-M. and Hudry-Clergeon, G., Oriented Fibrin Gels Formed by Polymerization in Strong Magnetic Fields, *Nature* **289**, 91-93 (1981).

[4]Yamagishi, A., Takeuchi, T., Higashi, T. and Date, M., Diamagnetic Orientation of Polymerized Molecules under High Magnetic Field, *J. Phys. Soc. Jpn.* **58**, 2280-2283 (1989).

[5]Fujiwara, M., Fukui, M. and Tanimoto, Y., Magnetic Orientation of Benzophenone Crystals in Fields up to 80.0 KOe, *J. Phys. Chem. B* **103**, 2627-2630 (1999).

[6]Takahashi, T., Suzuki, K., Awano, H. and Yonetake, K., Alignment of Vapor-grown Carbon Fibers in Polymer under Magnetic Field, *Chem. Phys. Lett.* **463**, 378-382 (2007).

[7]Kimura, T., Study on the Effect of Magnetic Fields on Polymeric Materials and Its Application, *Polymer J.* **35**, 823-843 (2003).

[8]Schoenberg, D. and Uddin, M., The magnetic Properties of Bisumth I – Dependence of Susceptibility on Temperature and Addition of Other Elements, *Proc. Roy. Soc. A***156**, 687-701 (1936).

[9]Kohama, T. and Iwai, K., Continuous Measurement of XRD Peak Intensity of Bismuth under Magnetic Field, *J. Phys.: Conf. Ser.* **156**, 012003 (2009).

[10]Wang, Y. and Kim, K. S., Large-scale Polyol Synthesis of Single-crystal Bismuth Nanowires and the Role of NaOH in the Synthesis Process, *Nanotechnology* **19**, 265303 (2008).

[11]Watanabe, K., Watanabe, Y., Awaji, S, Fujiwara, M., Kobayashi, N. and Hasebe, T., X-ray Diffraction Investigation in High Fields at Low Temperature for $Nd_{0.5}Sr_{0.5}MnO_3$, *Advances in Cryogenic Engineering* **44**, 747-752 (1998).

[12]Kitamura, N., Takahashi, K., Mogi, I., Fukumi, K., Awaji, S. and Watanabe, K., Bismuth Nano-particle Dispersed Organic Composite for Optical Component, *Proc. SPIE* **8621** (2013) in print.

[13]Atkinson, R. and Lissberger, P. H., Optical Constant of Thin Film Bismuth, *Thin Solid Films* **17**, 207-221 (1973).

[14]Lenham, A. P., Treherne, D. M. and Metcalfe, R. J., Optical Properties of Antimony and Bismuth Crystals, *J. Opt. Soc. Am.* **55**, 1072-1074 (1965).

[15]Khashan, M. A. and Nassif, A. Y., Dispersion of the Optical Constants of Quartz and Polymethyl Methacrylate Glasses in a Wide Spectral Range: 0.2-3 μm, *Opt. Commun.* **188**, 129-139 (2001).

CONTROL OF DISPERSION AND AGGLOMERATION OF CNTS FOR THEIR NETWORKING - MECHANICAL AND ELECTRICAL PROPERTIES OF CNT/ALUMINA COMPOSITES -

Mitsuaki Matsuoka, Junichi Tatami, Toru Wakihara
Graduate School of Environment and Information Sciences, Yokohama National University
79-7 Tokiwadai, Yokohama, 240-8501, Japan

ABSTRACT

Carbon nanotubes (CNTs) are expected to be used to improve electrical conductivity and to reinforce mechanical properties of many materials. In our previous study, fabrication of CNT-dispersed Si_3N_4 ceramics having high strength succeeded by uniform dispersion of CNTs using bead milling technique. However, their electrical conductivity degraded though the size of CNTs was not changed after bead milling process. The reason for the degradation is that formation of electrically conductive paths due to networking of CNTs was insufficient. In this study, networking of CNTs in ethanol was controlled by adding surfactant. Furthermore, CNT/alumina composites were fabricated by spark plasma sintering and their electrical and mechanical properties were evaluated. From optical microscopic observation, networking of CNTs in the slurry was occurred by adding ammonium polycarboxylate. Electrical conductivity of CNTs dispersed slurry increased with an increase in quantity of the added ammonium polycarboxylate and with an increase in the soaking time due to development of CNT network. Electrical conductivity of the CNT/alumina composites prepared by adding ammonium polycarboxylate were higher than those without ammonium polycarboxylate while having higher strength. The improvement of these properties resulted from homogeneous networking of CNTs formed by adding ammonium polycarboxylate to increase the conduction paths in the sintered body.

INTRODUCTION

Carbon nanotubes (CNTs) have high electrical conductivity, high strength and so on. Improvement of electrical conductivity and reinforcement of mechanical properties by adding CNTs are expected in many kinds of materials. In our previous study, fabrication of CNT-dispersed Si_3N_4 ceramics having high strength succeeded by uniform dispersion of CNTs using bead milling technique. However, their electrical conductivity degraded compared with those of ball milling though the size of CNTs was not changed after bead milling process. The reason for the degradation of electrical conductivity was that formation of electrically conductive paths due to networking of CNTs was insufficient. In this study, networking of CNTs in ethanol was controlled by adding surfactant. Furthermore, CNT/alumina composites were fabricated by spark plasma sintering to improve the mechanical and electrical properties simultaneously.

EXPERIMENTAL PROCEDURE

Multiwall carbon nanotubes (VGCFs, Showa Denko Co., Japan) having an average diameter of 60 nm and an average length of 6 μm were used in this study. As-received CNTs formed strong agglomerates or bundles by van der waals forces. The CNTs were dispersed in ethanol by bead milling. First, 0.2 wt% CNTs were added to ethanol with polyethyleneimine (PEI) having a molecular weight of 250 as a cationic surfactant. The amount of the added PEI was five times as much as the weight of CNTs. After ultrasonic irradiation for 20 min. Bead milling (Minicer, Ashizawa Finetech Ltd., Japan) were applied to the dispersed CNTs in ethanol for 2 h at 3000 rpm. Al_2O_3 beads 300 μm in diameter were used as grinding media. In order to control networking of CNTs, ammonium polycarboxylate which is the anionic surfactant having the opposite charge of PEI was added to CNT-dispersed slurry.

Aluminum isopropoxide (AIP) was added to the CNT-dispersed slurry as a precursor of Al_2O_3. After making networking of CNTs in the AIP solution by adding PCA, water were added to them to form moistening gels of $Al(OH)_3$. The moistening gels were dried for 24 h at 80°C to make dry gels of Al_2O_3. They were sieved to obtain granules using a 1000 μm mesh sieve. After dewaxing, the granules were put into graphite mold and were sintered for 1 min at 1300~1400°C in Ar under 80 MPa by spark plasma sintering process.

RESULTS AND DISCUSSION

Figure 1 shows the optical micrographs of CNT-dispersed slurry before and after adding ammonium polycarboxylate. CNTs was dispersed well homogeneously by using bead milling process, and there is not agglomeration of CNTs in Fig. 1 (a). Brownian motion of CNTs was also observed. After adding ammonium polycarboxylate, networking of CNTs in the slurry was occurred in Fig. 1(b). It was observed that networking of CNTs in the slurry was three dimensionally formed because of adding opposite charge surfactant. Less than or equal to the 1.0wt% ammonium polycarboxylate addition, the homogeneous network structure of CNTs was formed from homogeneously dispersion of CNTs, and it was developed with an increase of the added ammonium polycarboxylate. On the other hand, in the case of adding ammonium polycarboxylate more than 2.0wt%, the network of CNTs was changed into the cluster structure.

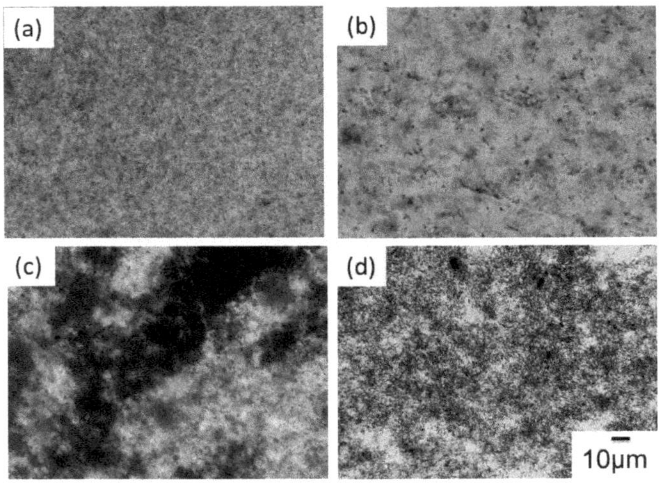

Figure 1. Optical micrographs of CNTs agglomeration by adding various amount of ammonium polycarboxylate.
(a)0 wt%, (b)0.4 wt%, (c)1.0 wt%, (d)2.0 wt%.

Figure 2 shows electrical conductivity of CNT-dispersed slurry by adding various amount of ammonium polycarboxylate. Electrical conductivity of CNTs dispersed slurry increased with an increase of the added ammonium polycarboxylate. When 1.0wt% ammonium polycarboxylate added (=network structure of CNTs formed), it reached a maximum value. Furthermore, the excess amount of ammonium polycarboxylate resulted in decreasing the electrical conductivity.

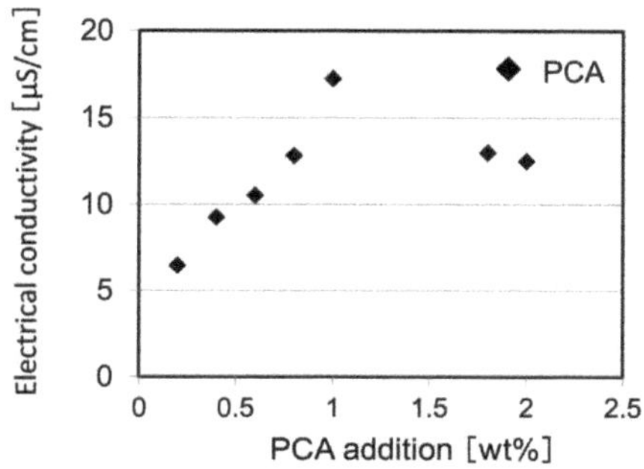

Figure 2. Electrical conductivity of CNT-dispersed slurry by adding various amount of ammonium polycarboxylate.

In the case of less than or equal to the 1.0wt% ammonium polycarboxylate addition, the electrical conductivity increased in proportion to the amount of ammonium polycarboxylate addition, and it also increased with time in Figure 3. When the added ammonium polycarboxylate was more than the 1.0wt%, the electrical conductivity decreased with the soaking time which was corresponded to the tendency of network structure. The optimum amount of the surfactant added, CNTs does not agglomerate after network formation.

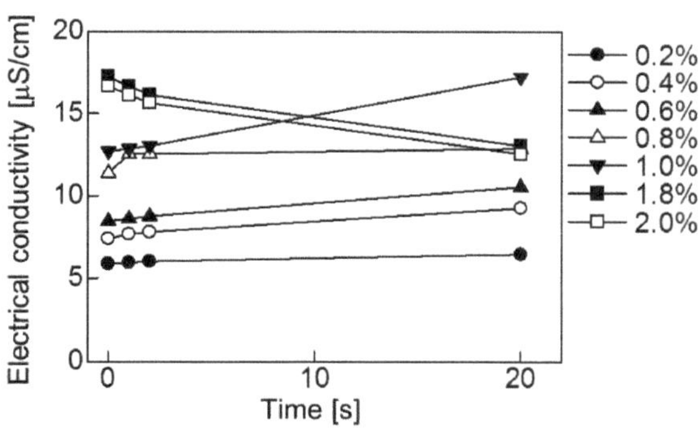

Figure 3. Time-dependence of electrical conductivity by adding various amount of ammonium polycarboxylate.

Table I shows properties of the CNT/alumina composites sintered by spark plasma sintering. The relative dinsities of all the specimens were higher than 96%. The bending strength of CNT/alumina composite prepared by adding PCA were as high as those of without ammonium polycarboxylate. The electrical conductivity of the CNT/alumina composite prepared by adding ammonium polycarboxylate were higher than those without ammonium polycarboxylate. The improvement of these properties resulted from homogeneous networking of CNTs formed by adding ammonium polycarboxylate to increase the conduction paths in the sintered body.

Table I. Properties of CNT/alumina composites (SPS)

CNT/alumina	Relative density /%	Bending strength / MPa	Electrical conductivity $/Sm^{-1}$
with PCA	96.2	622	532
without PCA	96.5	633	189

Figure 4 shows infrared micrographs of CNT/alumina composites with or without ammonium polycarboxylate. There were not CNTs agglomeration in the both specimens. The inner structure of the CNT/alumina composite with or without ammonium polycarboxylate were uniformity.

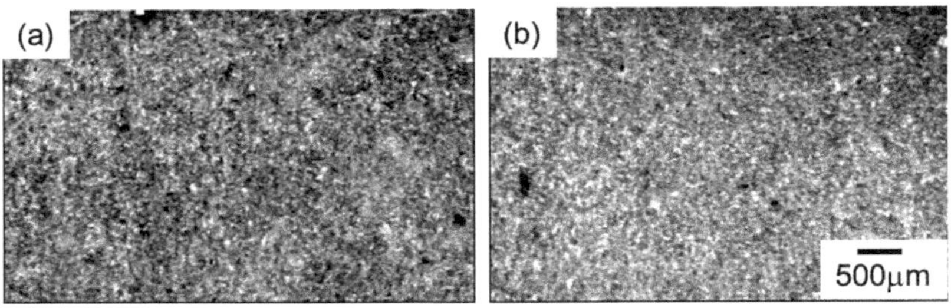

Figure 4. Infrared micrographs of CNT/alumina composites (a)without ammonium polycarboxylate, (b)with ammonium polycarboxylate.

CONCLUSIONS

Networking of CNTs in the slurry was formed by adding ammonium polycarboxylate. Electrical conductivity of CNT-dispersed slurry increased with an increase of the added ammonium polycarboxylate. When excess quantity of ammonium polycarboxylate was added, the network structure of CNTs was changed into the cluster structure to degrade the electrical conductivity. Network of CNTs was formed in CNT/alumina composites using CNT-dispersed slurry prepared by adding ammonium polycarboxylate. Electrical conductivity of the CNT/alumina composites prepared by adding ammonium polycarboxylate were higher than those without ammonium polycarboxylate while having higher strength. The improvement of these properties resulted from homogeneous networking of CNTs formed by adding ammonium polycarboxylate to increase the conduction paths in the sintered body.

REFERENCES

[1]J. Tatami, T. Katashima, K. Komeya, T. Meguro and T. Wakihara, "Electrically Conductive CNT-Dispersed Silicon Nitride Ceramics," *J. Am. Ceram. Soc.*, 88, 2889-2893 (2005).

[2]M. Matsuoka, S. Yoshio, T. Yamakawa, J. Tatami, T. Wakihara, K. Komeya, T. Meguro, "Development of CNT-Si_3N_4 Composites with High Strength and Electrical Conductivity by Adding HfO_2", Trans. MRS-J., 37[1] 11-14(2012).

[3]M. Matsuoka, S. Yoshio, J. Tatami, T. Wakihara, K. Komeya, T. Meguro, "Development of CNT-Si_3N_4 Ceramics by adding Lower Temperature Sintering Aids", J. Kor. Ceram. Soc., 49 [4] 333-336 (2012).

[4]S. Yoshio, J. Tatami, T. Yamakawa T. Wakihara, K. Komeya, T. Meguro, K. Aramaki and K. Yasuda, "Dispersion of Carbon Nanotubes in Ethanol by a Bead Milling Process," *Carbon.*, [26] 4131-4137 (2011).

SYNTHESIS AND MICROSTRUCTURE DEVELOPMENT IN YTTRIA-MAGNESIA CERAMICS FOR INFRARED TRANSPARENCY

J. A. Miller and I. E. Reimanis
Colorado Center for Advanced Ceramics
Metallurgical and Materials Engineering Department
Colorado School of Mines
Golden, Colorado, United States

ABSTRACT

Pure Y_2O_3 and varying compositions of $MgO-Y_2O_3$ were synthesized using thermal decomposition synthesis and compared to a commercially available powder. Specimens with densities up to 96% were made by hot pressing. The hardness of these specimens ranged from 6.3 GPa for the pure Y_2O_3 to 10.4 GPa for the commercially available powder. The transmission of wavelengths from 2,500 nm to 25,000 nm was measured. The specimens made with the in house synthesized powders achieved maximum transmissions of approximately 1.9%, 12.1%, and 0.3% in the 0-100 vol.%, 25-75 vol.%, and 50-50 vol.% $MgO-Y_2O_3$ specimens, respectively, while a maximum transmission of approximately 45.5% was measured with the commercial specimen. Imaging of the microstructure reveals distinct MgO and/or Y_2O_3 phases in all the samples. The MgO and Y_2O_3 rich regions of the in house synthesized specimens are on the order of microns, whereas these regions are only on the order of a hundred nanometers in size in the specimen made with commercial powder. Relations between microstructure, composition, hardness and transmissivity are discussed.

INTRODUCTION

Ceramics transparent in the infrared are desirable because they provide corrosion, erosion and impact resistance in applications like transparent armor and protective casings that shield infrared signal transmitters, such as radomes. $MgO-Y_2O_3$ nanocomposites represent a potential candidate system for these applications. Properties of ceramics for radomes are typically benchmarked against those of single crystals such as sapphire which exhibit high hardness and good optical transmission; sapphire has a Knoop hardness of 22 GPa and maintains 94% transmission to 2500 cm^{-1} (4,000nm). However, single crystal armor and radomes are expensive to produce, in part, because of cutting and polishing single domes from large boules of material[1]. Furthermore, the toughness of single crystals is relatively low.

Y_2O_3 and MgO as single phases exhibit high transmissivity over a broad range of the infrared spectrum (transmissions up to 79% out to approximately 1500 cm^{-1} (6,667 nm) and 85% out to 1400 cm^{-1}(7,143 nm), for Y_2O_3 and MgO respectively[1]), even as polycrystalline materials, but they are relatively soft (Knoop hardness of Y_2O_3 is 7.2 GPa; that for MgO is 6.4 GPa). Y_2O_3-MgO composites have been engineered with the goal to improve the mechanical properties while maintaining the high transmissivity[2-8]. A recent study reveals that an exceedingly high hardness of 23.1 GPa is attainable when the constituent grain size is kept to around 150 nm or less[9]. In another study, a hardness of 7.5 GPa was observed for a 50-50 vol.% composition synthesized by plasma spraying[10]. It is apparent that microstructure is critical in determining the hardness of $MgO-Y_2O_3$ composites.

The motivation for the present study is to understand the mechanical behavior of MgO-Y_2O_3 composites and the relation of microstructure with optical transmission. A variety of techniques have been employed for the synthesis of $MgO-Y_2O_3$ powders, ranging from sol-gel synthesis, precipitation from carbonates and hydroxides[12], and decomposition of nitrate-hydrates[13]. Muoto took the decomposition of nitrates a step further by adding an ammonium acetate fuel source to increase the rate of reaction during calcination to minimize the final grain size[14]. The present study used Muoto's process to synthesize $MgO-Y_2O_3$ powders in order to compare how hardness and transmissivity vary as the microstructure changes with varying amounts of MgO and Y_2O_3 phases present.

EXPERIMENTAL PROCEDURE

Pure Y_2O_3, 25-75 vol.% $MgO-Y_2O_3$, and 50-50 vol.% $MgO-Y_2O_3$ powders were created from thermal decomposition synthesis of ammonium acetate($C_2H_7NO_2$, Alfa-Aesar, 97%) w/ magnesium nitrate hexahydrate ($Mg(NO_3)_2{:}6H_2O$, Sigma Aldrich, ACS reagent 99%) and yttrium nitrate hexahydrate ($Y(NO_3)_3{:}6H_2O$, Alfa-Aesar, 99.9%, REO). The acetate and nitrate precursors were mixed in DI water with a magnetic stirrer for 15 minutes and then dried in a drying oven at 150°C for 24-48 hours. The dried powders were then calcined by heating at 2 C/min to 700°C and holding for 2 hours in a Deltech box furnace. X-ray diffraction was used (Phillips X'Pert Pro) to confirm that MgO and Y_2O_3 phases had been created from the nitrate precursors, and the powders were then hot pressed (Thermal Technology 2450) at 35 MPa to 1300°C for 1 hour in graphite die to produce 13 mm diameter samples varying from 4-7 mm thick. 50-50 vol.% $MgO-Y2O_3$ produced by Nanocerox via flame spray pyrolysis was hot pressed under identical conditions for comparison of the in house synthesized materials to commercially available materials.

The hot pressed specimens were ground to a uniform thickness of 3.5 mm using a surface grinder (Harig 618 Auto step) with a 400 grit diamond wheel. The samples were then cut into two semi-circular pieces; one piece of each specimen was polished to 1 μm finish and Vickers hardness was measured (Leco MHT Series 200). A load of 1000 g and a dwell time of 15 seconds was used for each measurement in accordance with ASTM standard C1327-08 (Standard test method for Vickers indentation hardness of advanced ceramics). Measurements were taken along the outer edge of the specimens.

The remaining half of each specimen was then further ground to a thickness of approximately 0.8 mm and polished to a 1 μm finish on one side and approximately 39 μm on the other side. Fourier transformed infrared spectroscopy (Nexus 670 FTIR E.S.P) was used to determine transparency from 400-4,000 cm^{-1} (wavelength 2,500-25,000 nm). In each case, the side of the sample with the 1 μm finish was faced towards the incident beam and the side with the 39 μm finish was faced towards the detector. Scanning electron microscopy (JEOL JSM-7000F SEM) was used to examine the microstructure of each specimen.

RESULTS AND DISCUSSION

X-ray diffraction (XRD) confirmed the presence of MgO and Y_2O_3 phases in the powders synthesized using the thermal decomposition process, as shown if Figure 1. All three of the synthesized powders showed peaks at the same 2θ with similar intensities. Additional yttrium oxide peaks were observed in the powders produced by Nanocerox. All samples showed similar diffraction peaks after hot pressing, and all peaks were identified as either MgO or Y_2O_3 for all four samples, although the relative intensities of the peaks varied according to the amounts of each phase present. All peaks in the 0-100vol.% $MgO-Y_2O_3$ (pure yttria) specimen were identified as Y_2O_3, as expected. Thus, XRD confirmed that the expected phases were present in all specimens, and no significant amount of unexpected phases were present post hot pressing, within the detection limits of XRD.

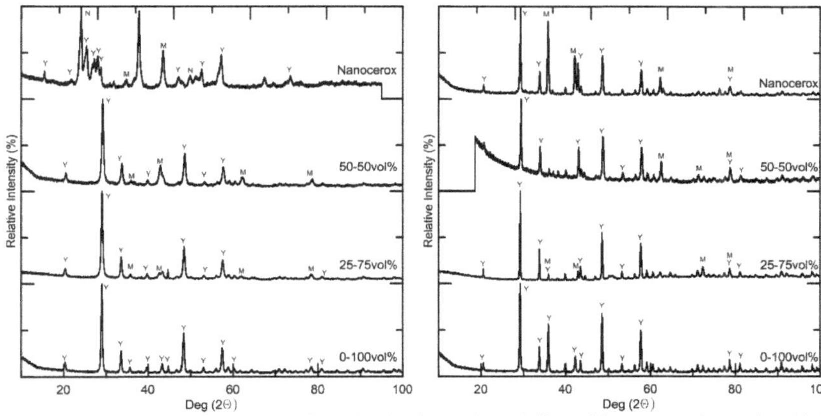

Figure 1. X-Ray diffraction peaks of synthesized powders (left) and hot pressed samples (right) for 0-100vol.% MgO-Y_2O_3 (pure yttria), 25-75vol.% MgO-Y_2O_3, 50-50vol.% MgO-Y_2O_3, and commercially available powder MgO-Y_2O_3. Y_2O_3 peaks are labeled Y and MgO peaks are labeled M. Note that an additional PDF (Yttrium-Oxide, 00-044-0399) was used to identify all of the peaks in the commercially available powder sample.

The hardness increased as the amount of MgO increased, with the exception of the synthesized 50-50vol.% specimen, as shown in Figure 2 and Table 1. Pure Y_2O_3 had an average indentation hardness of 640.1 HV (6.3 GPa) and a density of 96.2%, 25-75vol.% had a hardness of 808.7 HV (7.9 GPa) and a density of 96.2%, and the Nanocerox specimen had a hardness of 1055.9 HV (10.4 GPa) and a density of 96.2%. The synthesized 50-50vol.%, however, showed an average indentation hardness of 608.8 HV (6.0 GPa) and a noticeably lower density of 92.0%. It is assumed that the low hardness observed in the synthesized 50-50vol.% specimen is due to the low density.

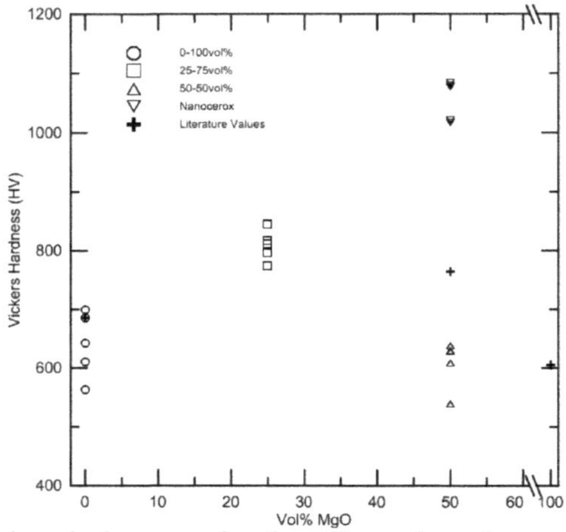

Figure 2. Vickers Hardness is shown as a function of increasing volume percent MgO present. Literature values for pure Y_2O_3, pure MgO, and 50-50vol.% MgO- Y_2O_3 are included for comparison.

Table I. Experimental Vickers Hardness data for 0-100vol.%, 25-75vol.%, 50-50vol.% MgO-Y_2O_3, and commercially produced Nanocerox MgO-Y_2O_3.

Material	Average VH	Std Dev	Density
0-100vol%	640.1	55.51	96.2%
25-75vol%	808.7	26.16	96.2%
50-50vol%	608.8	40.44	92.0%
Nanocerox	1055.9	33.44	96.3%

All specimens showed the highest transmission between approximately 3,330 nm and 10,000 nm. The specimen made using the commercial powder exhibited a maximum transmission of about 45 % at approximately 6,710 nm. The 0-100 vol.%, 25-75 vol.%, 50-50 vol.% specimens had maximum transmissions of 1 %, 12 %, and 0.3%, respectively at wavelengths of approximately 7,520 nm, 6,800 nm, and 8,400 nm, respectively. Transmission spectra from 2,500 nm to 25,000 nm are shown separately for each composition in Figure 3. In order to clearly show the transmission of each specimen, varying vertical axis limits are used. All spectra are shown on a common scale in Figure 4.

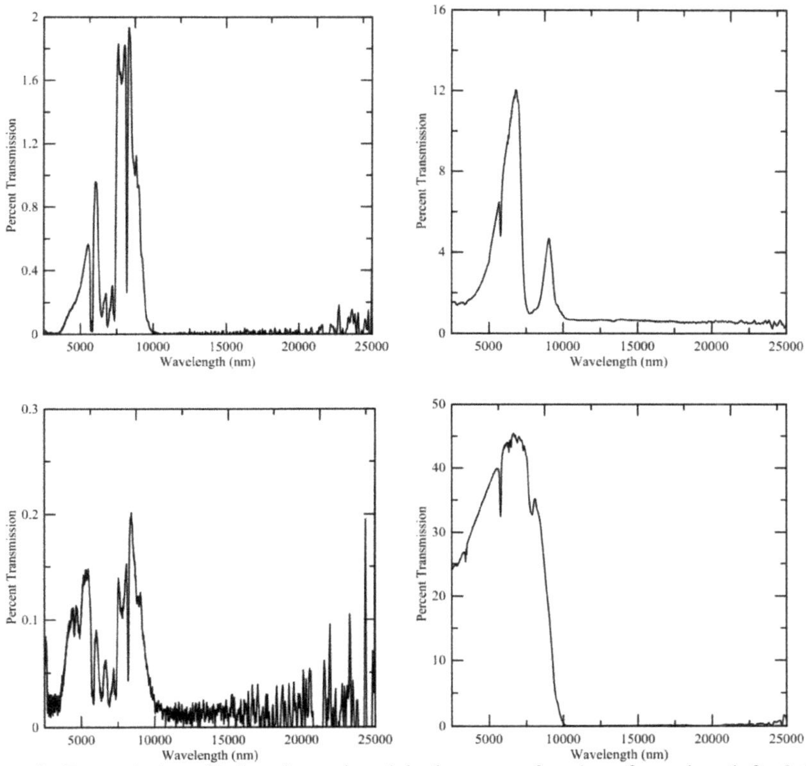

Figure 3. Transmission of infrared wavelength is shown as a function of wavelength for 0-100 vol.% (top left), 25-57 vol.% (top right), 50-50 vol.% (bottom left), and Nanocerox (bottom right).

Some similar absorption bands were observed in all specimens. Absorption from 3,330 nm and 3,640 nm is attributed to stretching O-H bonds likely due to moisture found within the FTIR chamber and specimen[15]. These absorption bands are seen most clearly in the spectra for the commercial and 25-75 vol.% materials. The absorption peaks observed at approximately 5,680 nm and 6,250 nm correspond to stretching C=O bonds of carboxylic acid groups of the form COOH[15]. Such compounds may be residual from the acetate fuel source used in the thermal decomposition reaction. However, because both of the absorption bands are also present in the commercial sample, it is likely that they are a result of carbon contamination from the graphite die used during hot pressing. Other absorption bands of note occur at the higher wavelengths. For MgO, an absorption band at 22,470 nm has been reported[16]. For Y_2O_3, bands at 21,692 23,641, 30,030, 30,769, and 33,333nm have been reported[16]. In general, the absorption observed at wavelengths less than 3,330 nm is due to electronic transitions, while the absorption observed at wavelengths greater than 10,000 nm are due to lattice vibrations[1].

Both the 0-100 vol.% and 50-50 vol.% specimens show additional absorption bands at approximately 6,850 nm and 7,350 nm. Further investigation is needed to identify the cause of these bands and explain why they are not observed in the other two specimens.

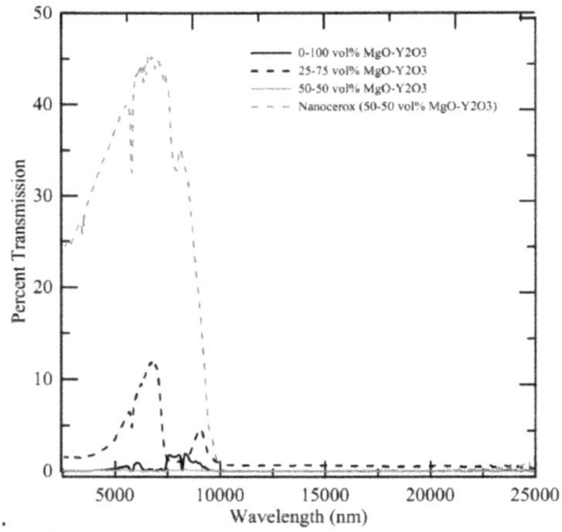

Figure 4. Transmission of 0-100 vol.%, 25-75 vol.%, 50-50 vol.%, and Nanocerox samples.

SEM images of the centers of the specimens show the difference structures in Figure 5. The in house synthesized specimens show a mixture of distinct Y_2O_3 (light) and MgO (dark) phases. In these samples, the size of each phase ranges from less than 1 μm to 2-3 μm, but individual grain size is not shown explicitly. The commercial specimen shows similar distinction between phases, but the phase regions are much better mixed and the size of each phase region is noticeably smaller.

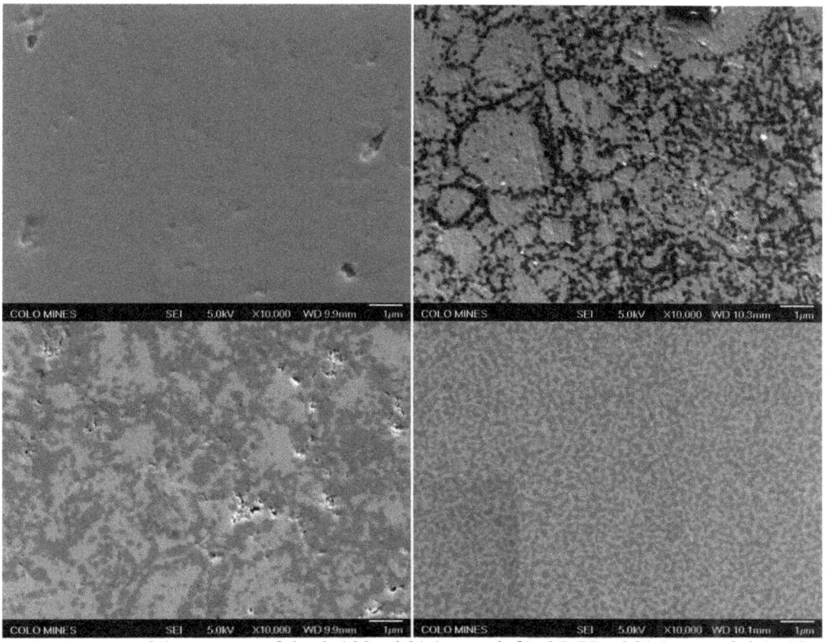

Figure 5. Images of the centers of the 0-100vol.% (upper left), 25-75vol.% (upper right), 50-50vol.% (lower left), and Nanocerox (lower right) samples taken at 10,000X

The differences in microstructure between the in-house synthesized specimens and the commercial material indicate that the size of the phase regions may impact the overall transmission of the material. The low overall transmission observed in the 50-50 vol.% specimen is attributed to scattering due to porosity as it has a notably lower density than the remaining specimens.

CONCLUSION

Hardness increased with the amount of MgO present, except for the synthesized 50-50vol.%. The low hardness of this sample is attributed to the low density. When infrared transmission of MgO-Y_2O_3 synthesized using thermal decomposition synthesis was compared to commercially available powders for wavelengths from 2,500 nm to 25,000 nm, it was found that these in-house powders had lower transmission, despite some compositions having similar densities. These differences are thought to be due to the differences in microstructure and the size of the phase regions within the material. Further work is needed to fully understand the fundamental mechanisms behind these differences so that future materials may bettered tailored to specific infrared transparent applications.

ACKNOWLEDGEMENTS

Colorado School of Mines and the authors thank Corning Incorporated for their financial support of this research.

REFERENCES

[1]D. C. Harris, *Materials for Infrared Windows and Domes, Properties and Performance.* Bellingham, Washington: SPIE- The International Society for Optical Engineering, 1999.
[2]G. Fantozzi, G. Orange, M. Gautier, J. Duraud, P. Maire, C. Le Gressus, and E. Gillet, "Effect of Nonstoichiometry on Fracture Toughness and Hardness of Yttrium Oxide Ceramics," *Journal of the American Ceramic Society*, vol. 72, no. 8, pp. 1988–1989, 1989.
[3]P. Hogan, T. Stefanik, C. Willingham, R. Gentilman, R. Integrated, and D. Systems, "Transparent Yttria for IR Windows and Domes – Past and Present," in *DoD Electromagnetic Windows Symposium*, 2004.
[4]D. Jiang and A. K. Mukherjee, "Synthesis of Y2O3-MgO nanopowder and infrared transmission of the sintered nanocomposite," *The International Society for Optics and Photonics*, vol. 7030, pp. 703007–703007–6, Aug. 2008.
[5]D. Jiang and A. K. Mukherjee, "The influence of oxygen vacancy on the optical transmission of an yttria–magnesia nanocomposite," *Scripta Materialia*, vol. 64, no. 12, pp. 1095–1097, Jun. 2011.
[6]M. J. Readey, R. Lee, J. W. Halloran, and A. H. Heuep, "Processing and Sintering of Ultrafine MgO-ZrO2 and (MgO, Y2O3)-ZrO2 Powders," *Journal of the American Ceramic Society*, vol. 73, no. 6, pp. 1499–1503, 1990.
[7] J. Wang, L. Zhang, D. Chen, E. H. Jordan, and M. Gell, "Y2O3-MgO-ZrO2 Infrared Transparent Ceramic Nanocomposites," *Journal of the American Ceramic Society*, vol. 1037, p. n/a–n/a, Nov. 2011.
[8] M. Jain, G. Skanda, A. Singhal, D. Agrawal, Y. Feng, and K. Olson, "Processing of nanopowders into transparent ceramics for infrared windows- Jain.pdf."
[9]C. S. Nodahl, T. Hartnett, T. Gattuso, R. Gentileman, and J. McCloy, "Mechanical, Optical, and Thermal Properties of Nano-Composite Optical Ceramics," Proceedings of the 12th DoD Electromagnetic Windows Symposium (2008).
[10] J. Wang, E. H. Jordan, and M. Gell, "Plasma Sprayed Dense MgO-Y2O3 Nanocomposite Coatings Using Sol-Gel Combustion Synthesized Powder," *Journal of Thermal Spray Technology*, vol. 19, no. 5, pp. 873–878, Apr. 2010.
[11]Jafar. F. Al-Sharab, B. H. Kear, S. Deutsch, and S.D. Tse, "Grain Size Refinement and Surface Modification of Bulk Polycrystalline Y_2O_3," in Proceedings of ICCE-18 Anchorage, (Alaska, USA July 4-10, 2010).
[12]T. Ikegami, J.-G. Li, T. Mori, and Y. Moriyoshi, "Fabrication of Transparent Yttria Ceramics by the Low-Temperature Synthesis of Yttrium Hydroxide," *Journal of the American Ceramic Society,* vol. 85, pp. 1725-1729, 2002.
[13]Article P. Melnikov, V. A. Nascimento, L. Z. Z. Consolo and A. F. Silva 2013 Mechanism of thermal decomposition of yttrium nitrate hexahydrate, Y(NO3)3·6H2O and modeling of intermediate oxynitrates Journal of Thermal Analysis and Calorimetry Springer Netherlands 111 1 115-119 2013/01/01 Mechanism of thermal decomposition of yttrium nitrate hexahydrate, Y(NO3)3·6H2O and modeling of intermediate oxynitrates J Therm Anal Calorim 1388-6150 Rare earthJ
[14]C. K. Muoto, E. H. Jordan, M. Gell, and M. Aindow, "Phase Homogeneity in Y2O3-MgO Nanocomposites Synthesized by Thermal Decomposition of Nitrate Precursors with Ammonium

Acetate Additions," *Journal of the American Ceramic Society*, vol. 94, no. 12, pp. 4207–4217, Dec. 2011.
[15]D. Niu, R. W. Ashcraft, and G. N. Parsons, "Water absorption and interface reactivity of yttrium oxide gate dielectrics on silicon," *Applied Physics Letters*, vol. 80, no. 19, p. 3575, 2002.
[16]N. T. Mcdevitt and W. L. Baun, "Infrared absorption study of metal oxides in the low frequency region (700-240 cm-1)," *Spectrochimca Acta*, vol. 20, no. 1960, pp. 799–808, 1964.

FABRICATION OF FLAKE-LIKE BOEHMITE/CERIA OR ZINC OXIDE COMPOSITES FOR UV SHIELD COATING

Seizo Obata
Gifu Prefectural Ceramics Research Institute
3-11 Hoshigadai
Tajimi 507-0811, Japan

Susumu Kawai, Michiyuki Yoshida and Osamu Sakurada
Faculty of Engineering, Gifu University
1-1 Yanagido
Gifu 501-1193, Japan

Kenji Kido
Kawai Lime Industry Co., Ltd.
2093 akasaka-cho
Ogaki 503-2291, Japan

ABSTRACT

It was discussed that flaked shaped particle with nano sized ceria (CeO_2) or zinc oxide (ZnO) synthesized by wetting process was possible to use as UV adsorption materials. Titanate aqueous solution (TNB-Lac) prepared by direct reaction of titanium tetra-n-butoxide (TNB) with lactic acid (Lac) in water shows characteristics of a polyanion and acts as a dispersant for preparing aqueous CeO_2 slurries. The isoelectric point (IEP) of CeO_2 particles shifted to lower pH with increasing concentration of TNB-Lac. Well-diesperesd CeO_2 slurries were obtained by modification of the surface changes on CeO_2 to adjust pH and addition of TNB-Lac concentration. On the other hand, the IEP of boehmite particles was at around pH 9.5. According to this result, the boehmite/CeO_2 or boehmite/ZnO pomposities could be synthesized by the electrostatic attraction between positively charged boehmite substrate and negatively charged CeO_2 or ZnO nanoparticles with 0.1 mmol/g TNB-Lac at around pH between 6 to 7. The synthesized particles were characterized by spectrophotometer and scanning electron microscopy. With boehmite/CeO_2 or boehmite/ZnO volume ratio 80/20, the boehmite was covered with CeO_2 or ZnO nanoparticles by microstructural observation. The silicon resin film including 5 mass% the synthesized composite with 20 vol% CeO_2 gave the equivalent UV shield ability, compared with the property of the film including nano-sized CeO_2.

INTRODUCTION

Ceria (CeO_2), titania (TiO_2) and zinc oxide (ZnO) are expected materials in glass coating, painting and cosmetics, because of excellent optical properties such as ultraviolet (UV) shielding abilities and visible transparency. Especially, CeO_2 and ZnO has high efficiency of UV shielding abilities in widely UV range (~ 400 nm). Enhancing that property of these materials, it was necessary that nano-sized particles of CeO_2 or ZnO were prepared. However, enhancement of the dispersion of these particles in the media is very difficult, because of the high specific surface area of particles.

For enhancing the dispersion of a nano-sized particles, good dispersant was absolutely imperative. Polyelectrolyte dispersant, such as polyacrylic and polymethaclic acids [1,2] as polyanions and polyethylamine [3] as polycations have dispering effects along with fluidizing effects, and are widely used. On the other hand, the effects of polymers that do not adsorb onto the particle surface have attracted a great deal of attention. Poly-L-lysine as a cationic polyelectrolyte[4] and nanoparticles of zirconium oxide[5] act as depletants. We have examined the effective use of some zirconium (IV) oxy-salts as fluidizing materials for ceramic slurries[6, 7], and developed a new direct-casting method based on aqueous-acidic high solid loading of alumina slurries stabilized with zirconium (IV) acetate[8].

Ohya et al. reported that very stable aqueous titanate solution could be prepared by direct reaction of titanium (IV) tetraisopropoxide ($Ti[OCH(CH_3)_2]_4$, TIP) with α-hydroxycardoxylic acids ($HOCR_1R_2$-CO_2H), such as lactic acid ($CH_3CH(OH)CO_2H$, Lac) in water, and that (004)-oriented anatase film with high refractive index of 2.54 was obtained from this stable titanate solutions *via* the spin coating method[9]. Furthermore, they observed that lactic acid worked as a very effective stabilizer for TIP to yield the stable aqueous solution of titanates even molar ratio of Lac/TIP = 1. These solutions are very valuable as starting materials for other titanium compounds, including titanium oxide, because they have higher titanium content and lower carbon dioxide emission on firing to titanium oxide. In addition, the titanate solutions are free from hazardous halogen, nitrogen, sulfur and metal ions other than titanium. Titanium and zirconium both belong to Group 4 in the periodic table because they have a similar valence shell that holds four electrons. In previous studies, we reported that titanate aqueous solution (TIP-Lac) prepared by direct reaction of TIP with Lac in water showed characteristics of a polyanion and acted an effective dispersant for preparing alumina and titania slurries [10, 11]. We also reported that titanium aqueous solution could be prepared by direct reaction of titanium(IV) tetra-n-butoxide ($Ti[O^nC_4H_9]_4$, TNB) with Lac , and a well-dispersed and fluidized barium titanate and aluminium tianate slurries could be obtained by the addition of the aqueous solution (TNB-Lac)[12, 13].

On the other hand, the covering ability was another important key factor in these applications such as glass coating, painting and cosmetics. Enhancing of the covering ability, plate-like or flake-like particles such as mica and talc were used. And the boehmite compound is found in coatings and cosmetics as additive and substrate, because of high orientation particles with plated shape. We have previously reported that boehmite synthesized by hydrothermal process had various morphologies of particles such as flaked or plated shapes[14, 15].

We noted that flaked shaped particle with nano-sized CeO_2 or ZnO synthesized by wetting process was possible to use as UV-shielding materials. In this study, the titanium aqueous solution (TNB-Lac) was selected as a dispersant for CeO_2 or ZnO particles. And the present study was performed to elucidate the optimum preparing conditions to obtain the flake-like boehmite with nano-sized CeO_2 or ZnO by the solution process with hetero-coagulation method.

EXPERIMENTAL

Materials

The boehmite powder was used in this study was flake-like boehmite (BMF-2, Kawai Lime Industry Co., Ltd. Ogaki, Japan) with an average particle size of 8-10 μm and aspect ratio of 80-100.

The commercial ceria powder (CeO_2)(Nippon Yttrium Co., Ltd., Omuta, Japan) or zinc oxide powder (ZnO)(Kanto Chemical Co., Inc., Tokyo, Japan) were used as starting materials, respectively. The nano-sized CeO_2 and ZnO powders were finely grinded in water by the bead milling (LMZ015, Ashizawa Finetech Ltd., Narashino, Japan) with 0.2 mm ZrO_2 balls for 6 h. The average particle size of CeO_2 and ZnO powder were 80 nm and 100 nm, respectively. Titanate aqueous solution as a dispersant was prepared by directly mixing titanium tetra-n-butoxide (TNB) with lactic acid (Lac) and water in air at room temperature[12]. An exothermic reaction occurred instantly to yield a solid white mass. The solid product may be considered titanium hydroxide produced by the hydrolysis reaction. The solid product dissolved gradually with stirring and yielded a clear solution within 24 h, followed by addition of a suitable amount of water to adjust the titanium concentration to 1.0 M, after ridded the prepared solution of n-butanol using a separating funnel. The molar ratio of Lac to TNB had a value of 1. This aqueous titanate solution was designated as TNB-Lac. The titanium concentration of synthesized TNB-Lac ($[Ti]_{TNB\text{-}Lac}$) was determined gravimetrically by drying in air at room temperature and ignition of the dried TNB-Lac to titanium oxide. The pH of the slurries was adjusted with tetramethyl ammonium hydroxide (TMAOH) and HNO_3.

Preparation of slurries

Slurries containing water, TNB-Lac and CeO_2 or ZnO, which were added to the slurries in fixed amounts on dry-weight basis (dwb) for these powders, were prepared by ball milling using 5mm ZrO_2 balls for 24 h at room temperature. Slurries containing water and boehmite, which were added to the slurries in fixed amounts on dwb for the powders, were prepared by stirring with magnetic stirrer for 24 h at room temperature. The water used to prepare all of the slurries was distilled and purified using a Milli-Q system (Milli-Q Plus, Millipore Co., Bedford, MA, USA).

Synthesis processing of the composite

Boehmite/CeO_2 or ZnO composite was synthesized using the solution chemical route as follows. The slurries with boehmite, CeO_2 and ZnO particles, which were adjusted to certain pH value by TMAOH and HNO_3, were simultaneously dropped into the boehmite slurries with a speed of about 10ml/min. The mixed slurry was stirred with magnetic stirrer for 2 h and the pH value was kept at a certain value throughout the reaction. After filtration of the slurries, the obtained particles were repeatedly washed with water at several times, and dried in an oven at 120 ◦C for 24 h.

Measurements

The rheological behaviors of the slurries were measured at 25°C using a controlled-stress rheometer (MARS II, Haake, Karlsruhe, Germany) with a parallel plate sensor 35 or 60 mm in diameter. The zeta potentials were measured by acoustophoretic spectrometry (DT-1200; Dispersion Technology, Inc., NY, USA) to determine the particle surface charge. Measurement of the zeta potential was carried out with 2 vol% suspensions at 25°C. Adsorption isotherms were determined from the amount of non-adosorped titnium in the supernatant of the slurry using an ICP-AES (ICPS-7500, Shimadzu Co., Kyoto, Japan). The supernatant solutions were withdrawn after the 1 vol% slurries had been centrifuged at 10,000 g (10,000 rpm) for 1 h. Sedimentation volume was measured after pouring 100 ml of each 1vol% CeO_2 or ZnO slurry into graduated cylinders, which was then sealed and left to stand for 1week prior to measurement.The UV-shielding ability of the prepared samples were evaluated by measuring the transmittance spectra of thin films with the sample powders using an UV–vis spectrophotometer (V-670, JASCO Corporation, Tokyo, Japan).After mixing of 0.1 g of the sample and 1.9 g of liquid silicon regin in a polyethylene beaker, prepared regin were degassed under vacuum and applied onto a glass plate with an applicator. The film prepared onto glass plate was solidified at 60°C for 1 h. The microstructure of sintered bodies was observed by scanning electron microscopy (SEM, JSM-7001GC, JEOL Ltd., Tokyo, Japan).

RESULTS AND DISCUSSION

Electrokinetic properties of CeO_2 particles

Figure 1 shows the electrokinetic behavior of aqueous CeO_2 slurries with various amounts of TNB-Lac as a function of pH. The zeta potential is strongly dependent on the pH of the slurry. The CeO_2 particles have an isoelectric point (IEP) at around pH 9.5 without TNB-Lac. At pH > 9.5, the particles exhibit a negative charge. As a result, the mobility values decrease as the pH values increase. As the pH value increased from 9.5 to 12, the absolute value of the zeta potential increased. This was attributed to the adsorption of the OH^- ions onto the particle surface, which enhanced the electrostatic repulsive force. In the wide ranges from neutral to acidic range, a similar phenomenon was observed but with the adsorption of positive H^+ ions in the slurry. As the pH value decreased from 9.5 to 3, the value of the zeta potential increased. However, with the addition of more H^+ ions to the slurries, i.e., when the pH value decreased from 3 to 2, the large number of positive ions resulted in a reduction of

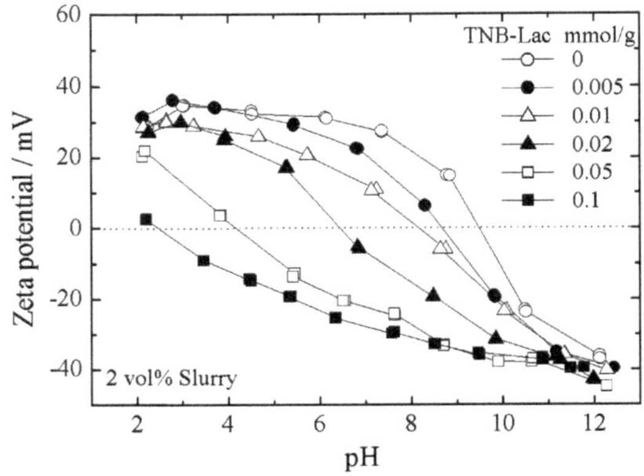

Fig. 1 Zeta potentials of CeO_2 particles with various amounts of TNB-Lac in the slurries as a function of pH.

the double layer thickness, and hence a reduction of the repulsive force between the particles. At acidic pH, the measured zeta potential appeared to be much lower in absolute magnitude than that at basic pH. On the other hand, examination of the zeta potential of CeO_2 slurries with TNB-Lac showed that the IEP of CeO_2 shifted to lower pH with increasing added amounts of TNB-Lac. These phenomena could be attributed to the adsorption of the negatively charged TNB-Lac as mentioned above on the surface of CeO_2 particles. The electrokinetic behaviors of aqueous ZnO slurries were similar to that of CeO_2 slurries, and the IEP of ZnO particle was around pH 9.2 without TNB-Lac. The IEP of ZnO shifted to lower pH with increasing added amounts of TNB-Lac. However, the zeta potentials of ZnO particles with various amounts of TNB-Lac were not measured at less than pH 6.5, because dissolution of ZnO was abruptly enhanced at less than pH 6.8[16].

Fluidity and dispensability of the CeO_2 slurry with TNB-Lac

Figure 2(a) shows the effect of the concentration of TNB-Lac on the apparent viscosity of 15 wt% CeO_2 slurries as a function of pH. The apparent viscosity value of the CeO_2 slurries without TNB-Lac increases abruptly up to pH 10, and has a maximum value at pH10, and above then slightly decreases with increasing at pH. In the CeO_2 slurry with 0.02 mmol/g TNB-Lac, the apparent viscosity of the slurry gradually increases up to pH 6, and then, gradually decreases at more than pH 6. Adding 0.1 mmol/g TNB-Lac in CeO_2 slurry, the apparent viscosity values of the slurries exhibits lower values at more than pH 6. These behaviors may be reflected in the sedimentation volume of the CeO_2 slurries left to stand for 1week as shown Fig. 2(b). The sedimentation volume of the CeO_2 slurries without TNB-Lac decreases abruptly up to pH 8, and has a minimum value at pH8-10, and above then increases with increasing at pH. In the CeO_2 slurry with 0.02 mmol/g TNB-Lac, the sedimentation volume of the slurry gradually decreases up to pH 6, and then, gradually increases at more than pH 8. Adding 0.1 mmol/g TNB-Lac in CeO_2 slurry, the sedimentation volume of the slurry exhibits higher values at more than pH 6. The IEP of the CeO_2 slurries added amounts of 0 mmol/g, 0.02 mmol/g and 0.1 mmol/g were at pH 9.5, pH 6.5 and pH 2.4, respectively, as shown Fig. 1. This observation indicated that dispersion of CeO_2 particles was controlled by addition of TNB-Lac.

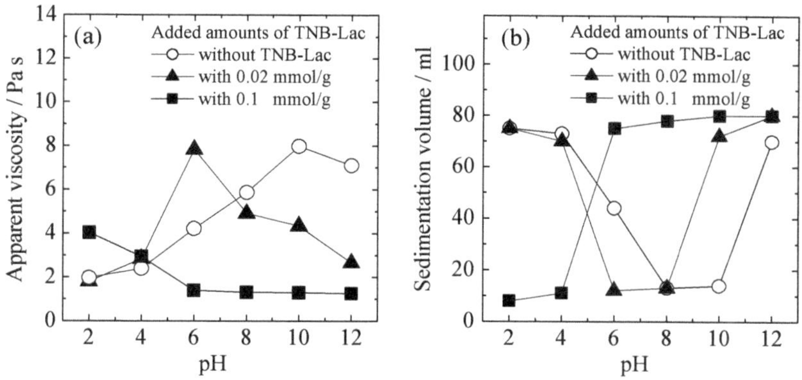

Fig. 2 (a) Apparent viscosities and (b) sedimentation volume of CeO_2 slurries with various amounts of TNB-Lac as a function of pH.

The adsorption of anionic TNB-Lac on the positively charged CeO_2 particles would reduce the surface charge of CeO_2, and charge neutralization could conduct agglomeration[1]. In further addition of TNB-Lac, CeO_2 also has the opposite surface charge to be re-dispersed.

Adsorption isotherm of TNB-Lac on CeO_2

Figure 3 shows the adsorption isotherm of TNB-Lac on CeO_2 surface as a function of the initial TNB-Lac concentrations at pH 10. A dotted line that presents complete adsorption (100% adsorption) of added TNB-Lac on the CeO_2 surface is also shown in the figure for comparison.
The adsorption was nearly 100% when the initial concentration was very low. The adsorbed amount of TNB-Lac increased with the increasing TNB-Lac concentration until a plateau (saturated adsorption) region was reached. The isotherm adsorption was 6×10^{-3} mmol m^{-2} when more than 0.1 mmol g^{-1} was used. This observation indicated the adsorption of anionic TNB-Lac on the CeO_2 surface.

Synthesis of boehmite/CeO_2 and boehmite/ZnO composites

Figure 4 shows that the zeta potential of boehmite, CeO_2 with 0.1 mmol/g TNB-Lac and ZnO with 0.1 mmol/g TNB-Lac in the slurries as a function of pH. The IEP of boehmite particles was at around pH 9.5. At pH > 9.5, the particles exhibit a negative charge. As the pH value decreased from 9.5 to 3, the value of the zeta potential increased. And, the zeta potential of boehmite had the behavior similar to the zeta potential of CeO_2. According to this result, the electrostatic attraction between positively charged boehmite substrate and negatively charged CeO_2 or ZnO nanoparticles with 0.1 mmol/g TNB-Lac could be expected at pH between 2.4 and 9.5. The boehmite/CeO_2 and boehmite/ZnO composite materials were synthesized by hetero coagulation reaction at around pH 6 and pH 7, respectively with boehmite/CeO_2 or boehmite/ZnO volume ratio 80/20. Figure 5 shows that the SEM photograph of the synthesized boehmite/CeO_2 composite and boehmite/ZnO composite. The image of showed that the boehmite was covered with CeO_2 nanoparticles uniformly. However, boehmite/ZnO composite was not well-synthesized than boehmite/CeO_2 composite. This result would be explained by the ion strength in the slurries because of the elution of ZnO into the media. In order to evaluate the properties of the prepared boehmite/CeO_2 composite and boehmite/ZnO composite, the

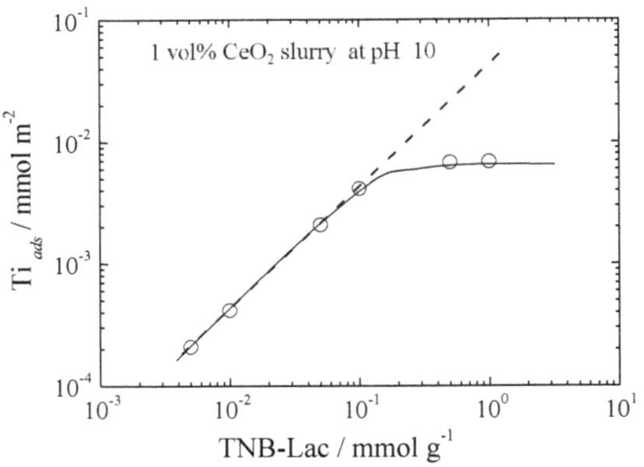

Fig. 3 Adsorption isotherm of TNB-Lac on CeO_2 surfaces as a function of the initial TNB-Lac concentrations.

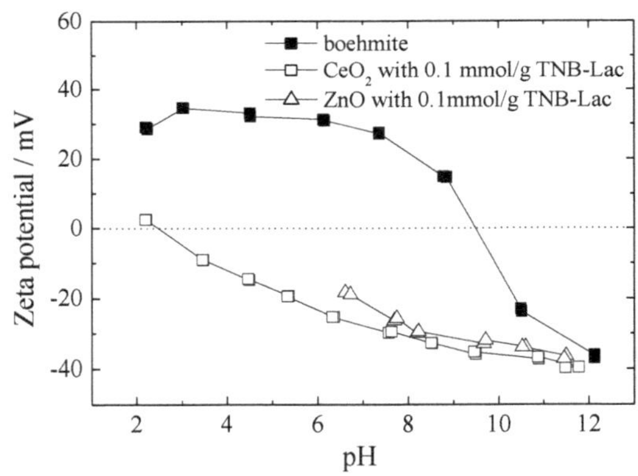

Fig. 4 Zeta potentials of boehmite, CeO_2 with 0.1 mmol/g TNB-Lac and ZnO with 0.1 mmol/g TNB-Lac in the slurries as a function of pH.

transmittance spectra have been carried out. Figure 6 shows the UV-vis transmittance spectra of thin films with/without 5wt% various composite particles. The transmittance in the visible light region of thin film with 5 wt% boehmite/CeO_2 had high value, and slightly decreased with the decreasing of wavelength. At 300 - 400 nm, the transmittance value abruptly decreased as the wavelength was decrease, and then the transmittance had lower value at less than 300 nm. This behavior was similar to the behavior of transmittance of thin film with CeO_2 particle. Especially, thin film including 5 mass%

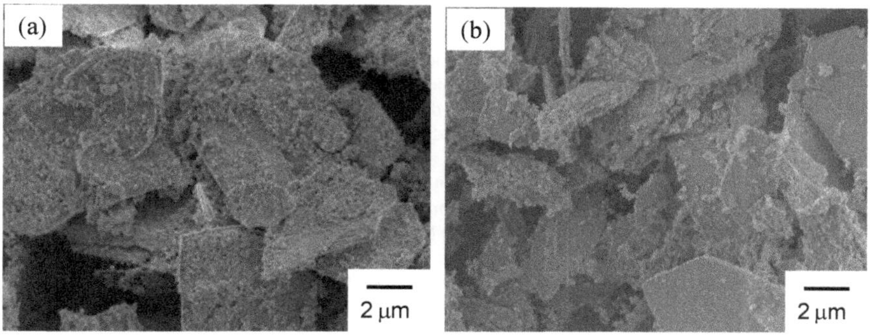

Fig. 5 SEM photograph of (a) boehmite/CeO_2 composite and (b) boehmite/ZnO composite: boehmite/CeO_2 or boehmite/ZnO volume ratio 80/20.

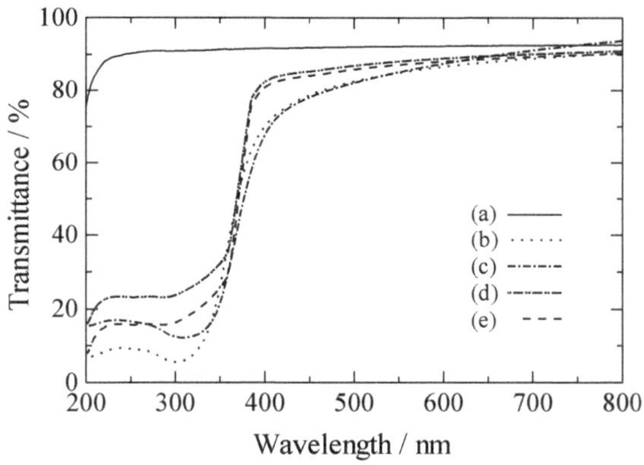

Fig. 6 UV-vis transmittance spectra of thin films (a) without or with (b) 5wt% CeO_2 (c) 5wt% ZnO, (d) 5wt% boehmite / CeO_2 composite and (e) 5wt% boehmite / ZnO composite.

the synthesized composite with 20 vol% CeO_2 gave the equivalent UV shield property, compared with the property of the film including nano-sized CeO_2. Unfortunately, in the case of behomite/ZnO composite, the transmittance was not obtain a favorable result, compared with that of the film including nano-sized ZnO, because boehmite was partially coated with nano-sized ZnO. In the further, we will report the processing for prepalation taking account of the elution of ZnO.

CONCLUSIONS

Titanate aqueous solution (TNB-Lac) prepared by direct reaction of titanium tetra-n-butoxide (TNB) with lactic acid (Lac) in water shows characteristics of a polyanion and acts as a dispersant for preparing aqueous ceria (CeO_2) slurries. The isoelectric point of CeO_2 particles shifted to lower pH with increasing concentration of TNB-Lac. Well-diesperesd CeO_2 slurries were obtained by modification of the surface changes on CeO_2 to adjust pH and control the TNB-Lac concentration. On the other hand, the IEP of boehmite particles was at around pH 9.5. According to this result, the boehmite/CeO_2 or boehmite/ZnO pomposities could be synthesized by the electrostatic attraction between positively charged boehmite substrate and negatively charged CeO_2 or ZnO nanoparticles with 0.1 mmol/g TNB-Lac at around pH between 6 to 7. The synthesized particles were characterized by spectrophotometer and scanning electron microscopy. With boehmite/CeO_2 or boehmite/ZnO volume ratio 80/20, the boehmite was covered with CeO_2 or ZnO nanoparticles by microstructural observation. The silicon resin film including 5 mass% the synthesized composite with 20 vol% CeO_2 gave the equivalent UV shield ability, compared with the property of the film including nano-sized CeO_2.

REFERENCES

[1] J. Cesarano III and I. A. Aksay, "Processing of Highly Concentrated Aqueous α-Alumina Suspensions Stabilized with Polyelectrolytes", *J. Am. Ceram. Soc.*, **71.** 1062-7 (1988).

[2] M. Itoh, O. Sakurada, M. Hashiba, K. Hiramatsu and Y. Nurishi, "Extensions of polyacrylic acid ammonium salts in the adsorption layer to fluidize alumina slurries", *J. Mater. Sci.*, **31.** 3321-4 (1996).

[3] F. Tang, T. Uchikoshi, K. Ozawa and Y. Sakka, "Dispersion of SiC Suspensions with Cationic Dispersant of Polyethylenimine", *J. Ceram. Soc. Japan*, **113.** 584-7 (2005).

[4] K. Furusawa, M. Ueda and T. Nashima, "Bridging and depletion flocculation of synthetic latices induced by polyelectrolytes", *Colloid Surf. A*, **153.** 575-81 (1999).

[5] V. Tohver, A. Chan, O. Sakurada and J. A. Lewis, "Nanoparticle Engineering of Complex Fluid Behavior", *Langmuir*, **17.** 8414-21 (2001).

[6] O. Sakurada, Y. Nakanishi and M. Hashiba, "Effect of zirconium acetate on the fluidity of acidic aqueous alumina slurries with high solids loading", *J. Mater. Sci. Lett.*, **20**. 929-31 (2001).

[7] O. Sakurada and M. Hashiba, "Depletion stabilization of ceramic suspensions with high solids loading in the presence of zirconium oxy-salts", *Stud. Surf. Sci. Catal.*, **132**. 375-8 (2001).

[8] N. Adachi, O. Sakurada and M. Hashiba, "Direct Coagulation Casting of Alumina Slurries Stabilized through Zirconium Acetate Using an Enzyme-Catalyzed Reaction", *Trans. Mater. Res. Soc. Jpn.*, **29**. 2037-40 (2004).

[9] T. Ohya, M. Ito, K. Yamada, T. Ban, Y. Ohya and Y. Takahashi, "Aqueous Titanate Sols from Ti Alkoxide-.ALPHA.-Hydroxycarboxylic Acid System and Preparation of Titania Films from the Sols", *J. Sol-Gel Sci. Tech.*, **30**. 71-81 (2004).

[10] O. Sakurada, M. Saito, T. Ohya, M. Hashiba and Y. Takahashi, "Dispersion and Fluidity of Aqueous Aluminum Oxide Suspension with Titanate Aqueous Solution", *J. Ceram. Soc. Japan*, **115**. 846-9 (2007).

[11] O. Sakurada, M. Komaba, S. Obata, M. Hashiba and Y. Takahashi, "Electrophoretic Deposition on Anodes from Aqueous Titania Suspensions with Titanate Solution", *Key Engineering Materials* **412.** 313-6 (2009).

[12] Japan Patent Publication No. 2007-161502.

[13] S. Obata, Y. Iwata, H. Yokoyama, O. Sakurada, M. Hashiba and Y. Takahashi, "Dispersion and Fluidity of Aqueous Aluminum Titanate Slurry by Addition of Titanate Aqueous Solution", *Ceramic Transactions* **212**, 227-34 (2010).
[14] Y. Kato, K. Kido, S. Obata, M. Hashiba and O. Sakurada, "Effect of pH on Hydrothermal Synthesis of Boehmite from Aluminum Hydroxide", 2009 Annual Meeting of The Ceram. Soc. Japan, 3H19 (2009).
[15] R. Miyazaki, M. Yoshida, K. Kido, S. Obata and O. Sakurada, "Morphologies of boehmite particles prepared by hydrothermal reaction with carbonate salts", 2011 Annual Meeting of The Ceram. Soc. Japan, 2G01 (2011).
[16] G. Charlot "Teiseibunsekikagaku", translated by K. Sone and M. Tanaka, pp297-298, Kyoritsu shuppan Co., Ltd., Japan (1974).

THERMAL DEGRADATION CONTROL STUDY OF CARBON FIBER / POLYAMIDE 6 COMPOSITE USING HEXAGONAL BORON NITRIDE POWDER

Daisuke Shimamoto*, Yusuke Imai and Yuji Hotta
National Institute of Advanced Industrial Science and Technology (AIST)
2266-98 Anagahora, Shimoshidami, Moriyama-ku, Nagoya, Aichi 463-8560, JAPAN
*Corresonding author

ABSTRACT

Thermal degradation of carbon fiber (CF) /thermoplastic composites (CFRTPs) is one of the problems because plastics are weak as for heating. Our developed carbon fiber/polyamide 6 (PA6) composites with hexagonal boron nitride (h-BN) particles (CF/h-BN/PA6) had higher thermal degradation resistance as compared to conventional carbon fiber/polyamide 6 (CF/PA6). When microwaves were irradiated to CF/h-BN/PA6 as thermal degradation test, PA6 matrix of one was not melting. In contrast, PA6 matrix of CF/PA6 was melted and decomposed by microwave irradiation. It was found that the improvement of thermal conductivity of PA6 matrix could inhibit thermal degradation of one by addition of h-BN.

INTRODUCTION

Recently carbon fiber/resin composites (CFRPs) are attracting attention as lightweight materials. CFRPs are expected to adapt to transportation equipment such as automobile and aircraft, structure material of wind power generator, building material etc.[1, 2] CFRPs are material that is composed carbon fibers (CFs) and plastics. CF has strong mechanical properties and thermal conductivity (10 W/m·K). On the other hand, thermoplastics (TPs) have easily fabrication properties by heating. TPs are attracting attention as CFRPs matrix. However, the thermal conductivity of TPs is much lower than CFs. When a lot of heat flows through to CF, thermal degradation[3] of TPs matrix occurs due to difference of thermal conductivity between CFs and TPs. It is problematic for the quality assurance of carbon fiber/thermoplastics (CFRTPs). If TP matrix of CFRTPs had more large thermal conductivity than existing TP matrix, thermal degradation of TP matrix would be inhibited with releasing heat to outside.

Hexagonal boron nitride (h-BN) is a ceramics powder whose structure is stacked plate-like particles[4] by van der Waals' force. Due to more than 150 W/m·K thermal conductivity,[5] h-BN is attracting attention as for thermal conductive filler. Moreover, their stacked structure reduces kneading machine damage because of low hardness.

In this study, improvement of thermal conductivity of plastic matrix was investigated by the addition and dispersion of the thermal conductive filler of h-BN to thermal degradation inhibition of CF/PA6 composite. Thermal degradation tests were carried out with microwave oven because microwave irradiation could selectively heat CFs in CFRTPs.

EXPERIMENTAL PROCEDURE

Materials

All chemicals and materials were used as received without further purification. Carbon fibers (CF, TORAYCA MLD-1000; average fiber length 150 μm) were obtained from Toray Industries, Inc.. Polyamide 6 (PA6, AMILAN CM1017) was used as a plastic matrix of CFRTPs that was purchased from Toray Industries, Inc.. Hexagonal boron nitride (h-BN, SHOBN UHP-1) was obtained from Showa Denko K. K. that showed insulation properties and lower dielectric loss than CFs.

Preparation of CFRTPs

CF and h-BN were mixed with PA6 using uniaxial kneading function (IMC-TAD3, Imoto Machinery Co., LTD.) for 20 rpm at 230 °C. The amount of CF and h-BN were adjusted to be 10 vol.%, respectively. After kneading, the pellets were pressed into 10 mm Φ disk at 230 °C. Disk shape test specimens were prepared about CF/PA6, h-BN/PA6 and CF/h-BN/PA6 composites.

Characterization

Thermal conductivity of CFRPs was investigated by the laser flash equipment (TC-7000, ULVAC, Inc.). The test specimens were prepared to be the size of 10 mm $\Phi \times 1$ mm t. Scanning electron microscopy (SEM, S4300, HITACHI Ltd.) measurement were performed at 10 kV to examine the nature of CF, h-BN and PA6 in CFRTPs. SEM samples were sputter-coated Pt-Pb. Apparent density of CFRTPs was estimated by Archimedian method using ethanol as a solvent (AX204, METTLER TOLEDO International Inc.). Thermal degradation test was carried out using microwave oven at 1.6kW power 2.45 GHz frequency (EM-1605, SANYO Electric Co., Ltd.) because microwave irradiation could selectively heat CFs in CFRTPs. The thermal degradation test specimen was prepared to be the size of 10 mm $\Phi \times 1$ mm t that is the same shape as for thermal conductivity measurement.

RESULTS AND DISCUSSION

Figure 1(a) shows that optical image of (left) the microwave irradiated CF/PA6 for 15 sec, (center) CF/PA6 for 30 sec and (right) CF/h-BN/PA6 for 60 sec. The microwave irradiated CF/PA6 for 15 sec was not melting. However, central part of the microwave irradiated CF/PA6 for 30 sec was melting. On the other hand, microwave irradiated CF/h-BN/PA6 for 60 sec was not melting. Irradiated microwaves were absorbed by CFs in CFRTPs and directly transformed into thermal energy.[5] As a result, CFs in CFRTPs were heated above melting temperature of PA6. Figure 1(b-d) shows fracture cross-section SEM images of CFRTPs, (b) non-microwave irradiated CF/h-BN/PA6, (c) microwave irradiated CF/PA6 for 15 sec and (d) microwave irradiated CF/h-BN/PA6 for 60 sec, respectively. h-BN

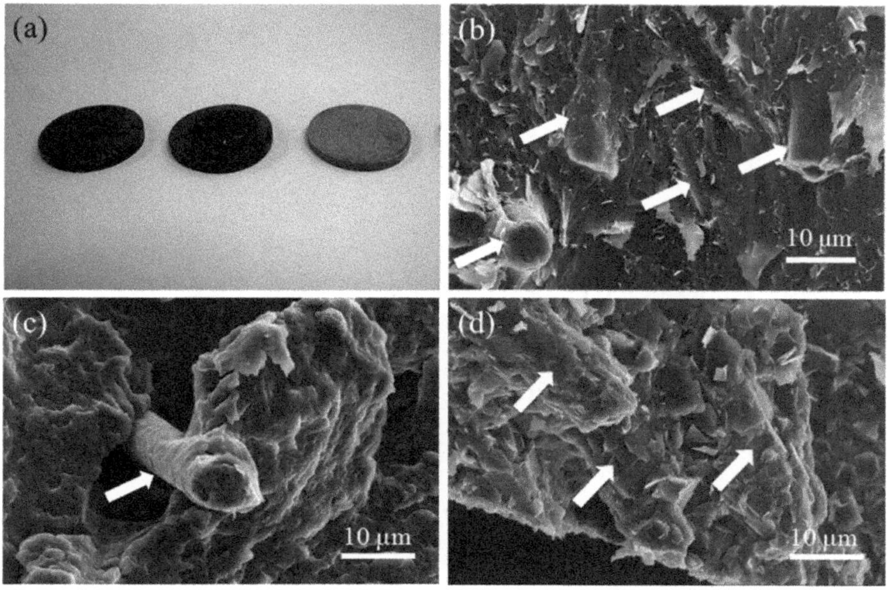

Figure 1 (a) Optical image of (light) microwave irradiated CF/PA6 for 15 sec, (center) CF/PA6 for 30 sec and (right) CF/h-BN/PA6 for 60 sec. Fracture cross-section SEM image of (b) un-microwave irradiated CF/h-BN/PA6, (c) microwave irradiated CF/PA6 for 15 sec and (d) microwave irradiated CF/h-BN/PA6 for 60 sec. White arrows shows CFs in CFRTPs.

in CF/h-BN/PA6 was dispersed in PA6 matrix as bridge between CF and CF (Figure 1(b)). Although optical image (Figure 1(a)) shows that microwave irradiated CF/PA6 for 15 sec is not melting, figure 1(c) shows thermal decomposition of PA6 around CF in the microwave irradiated CF/PA6 for 15 sec. The surface of CF formed facet structure. This means graphitization of CF. On the other hand, in more long time microwave irradiation (60 sec) for CF/h-BN/PA6, thermal degradation could not be confirmed (Figure 1(d)). CFs was covered with h-BNs and PA6 matrix. These results indicate that the addition of h-BNs reduces thermal degradation of PA6 matrix.

In order to clear the effect of h-BNs addition to CFRTP, thermal conductivity was estimated using laser flash method. Figure 2 shows thermal conductivity of CFRTPs in un-microwave and microwave irradiation. Un-microwave irradiated CF/PA6, h-BN/PA6 and h-BN/PA6 had clearly large thermal conductivity compared to PA6, which was to be 0.57, 0.67 and 0.96 W/m·K, respectively. The

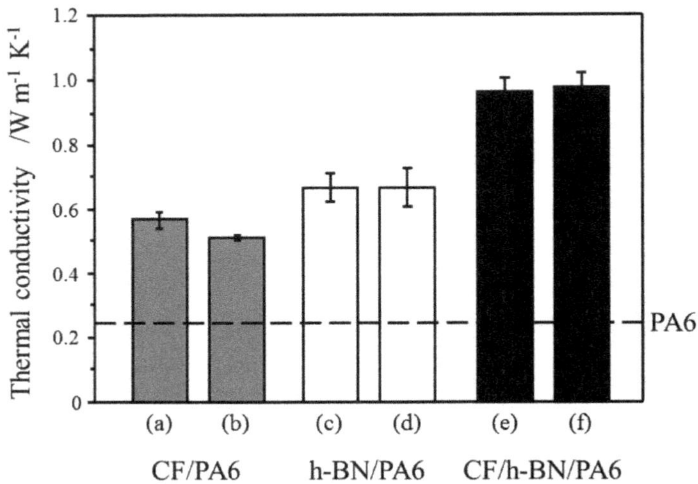

Figure 2 Thermal conductivity of the (a) un-microwave and (b) microwave irradiated CF/PA6 for 15 sec, (c) un-microwave and (d) microwave irradiated h-BN/PA6 for 60 sec, (e) un-microwave and (f) microwave irradiated CF/h-BN/PA6 for 60 sec.

thermal conductivity of CF/h-BN/PA6 was up 40 % from conventional CF/PA6 (Figure 2(a, e)). Because added h-BNs forms thermal conductive pass between CF and CF (see Figure 1(b)). In the un-microwave and microwave irradiation, h-BN/PA6 and CF/h-BN/PA6 showed similar thermal conductivity, respectively (Figure 2(c-f)). h-BN/PA6 does not absorb most of the microwaves due to low dielectric loss. That is to say, h-BN/PA6 is almost not heated by microwave irradiation. In case of CF/h-BN/PA6, CFs absorbs microwaves and produces heat. However CF/h-BN/PA6 can transfer heat to the whole through thermal conductive PA6 and CFs. Therefore, microwave irradiated CF/h-BN/PA6 are small damaged. Thus, thermal conductivity of microwave irradiated CF/h-BN/PA6 is similar to un-microwave irradiated one. In contrast, it is difficult for the CF/PA6 to transfer heat to the whole through low thermal conductive PA6. As a result, PA6 around CFs are decomposed and voids are produced (see Figure 1(c)). Therefore, in the microwave irradiated CF/PA6 for 15 sec, the decreasing of thermal conductivity from 0.57 to 0.51 W/m·K can be explained by decomposition for PA6.

Figure 3 shows the apparent density of the CF/PA6, h-BN/PA6 and CF/h-BN/PA6 in un-microwave and microwave irradiation for 60 sec. Apparent density of CF/P6 of un-microwave irradiation was to be 1.20 kg/m^3. The microwave irradiated CF/PA6 for 60 sec clearly showed low

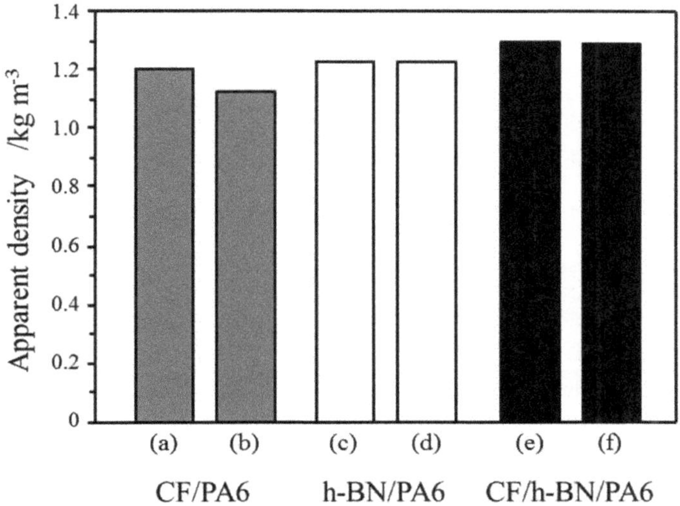

Figure 3 Apparent density of the (a) un-microwave and (b) microwave irradiated CF/PA6, (c) un-microwave and (d) microwave irradiated h-BN/PA6, (e) un-microwave and (f) microwave irradiated CF/h-BN/PA6 for 60 sec, respectively.

apparent density as compared to the un-microwave irradiated CF/PA6, which was to be 1.13 kg/m^3. PA6 in microwave irradiated CF/PA6 are thermally decomposed because of low thermal conductivity of PA6 matrix. On the other hand, the apparent density of the microwave irradiated h-BN/PA6 and CF/h-BN/PA6 for 60 sec were similar to the each un-microwave irradiated one. h-BN/PA6 is almost not heated by microwave irradiation due to low dielectric loss. In case of CF/h-BN/PA6, PA6 matrix is small damaged because of high thermal conductivity. Thus, improvement of thermal conductivity can inhibit thermal degradation of PA6 matrix.

Figure 4 shows SEM image of as received CF. The cross-section of CF is slightly-misshapen circular and the side surface of CF is curve line. In general, it is known that graphitized CFs have a characteristic polygonal appearance due to advanced graphite structure. Thus, the cylindrical form of CF becomes faceted form.[7] According to the result of Figure 1(c), microwave-heated CFs in CF/PA6 has polygonal side surface. Yoshida *et. al.* have been reported that graphitization occurs around at 2,500 °C.[8] In other word, microwave irradiated CFs in conventional CF/PA6 were heated the graphitization

Figure 4 SEM image of as received CF.

temperature. This temperature is enough high temperature to PA6 melt and pyrolyze. Thermal conductivity of PA6 is quiet low as compared to CF and h-BN. As a result, PA6 around CFs in CFRTP is decomposed into low hydrocarbon, carbon dioxide and water.[9] Thus, thermal decomposition of PA6 agrees with decreasing of apparent density. In contrast, it is found that h-BN powder is useful compound as inhibition thermal degradation for PA6. CF/h-BN/CF has a release heat property to the outside due to thermal conductive PA6 matrix. Therefore, CF/h-BN/CF could inhibit thermal degradation.

CONCLUSION

In this study, we found that the addition of h-BN could inhibit thermal degradation of conventional CFRTP (CF/PA6). h-BN of thermal conductive ceramics improved thermal conductivity of PA6 matrix in CFRTPs. In addition, microwave irradiated CFs was heated melting and pyrolysis temperature of PA6 matrix. Thermal conductive thermoplastics have effect in the thermal degradation of the plastic matrix for CFRTPs.

REFERENCES

[1] K.D.V.P. Yarlagadda and S. Hsu, *J. Mat. Proc. Tech.*, **155-156**, 1532-1538 (2004).

[2] R.A. Witik, F. Gaille, R. Teuscher, H. Ringwald, V. Michaud and J.E. Månson, *J. Cleaner Product.*, **29-30**, 91-102 (2012).

[3] S. Ray and R.P. Cooney, *Handbook ofEnviromental Degradation of Materials*, Elsevier Inc. 231-235 (2012).

[4] K. Sato, H. Horibe, T. Shirai, Y. Hotta, H. Nakano, H.Nagai, K. Mitsuishi and K. Watari, *J. Mat. Chem.*, **20**, 2749-2752 (2010).
[5] K. Watari and S.L. Shinde, *MRS Bull.*, **26**, 440-441 (2001).
[6] J.A. Menéndez, A. Arenillas, B. Fidalgo, Y. Fernández, L. Zubizarreta, E.G. Calvo, J.M. Bermúdez, *Fuel Proc. Tech.*, **91**, 1-8 (2010).
[7] T. Koyama, M. Endo and Y. Hishiyama, *Jap J. Appl. Phys.*, **13**, 1933-1939 (1974).
[8] A. Yoshida, Y. Hishiyama, M. Ishioka and M. Inagaki, *TANSO*, **168**, 169-175 (1995).
[9] B.G. Kim and D.G. Lee, *Nanotechnology*, **21**, 185305 (2010).

SOL-GEL AUTO-COMBUSTION SYNTHESIS OF CO-DOPED ZNO DILUTED MAGNETIC SEMICONDUCTOR NANOPOWDERS

Chuanbin Wang, Xuan Zhou, Fei Chen, Qiang Shen, Lianmeng Zhang
State Key Laboratory of Advanced Technology for Materials Synthesis and Processing, Wuhan University of Technology, Wuhan 430070, P. R. China

ABSTRACT

Co-doped ZnO diluted magnetic semiconductors show the special functionalities in the areas of magnetic, magneto-optical or magneto-electric devices. In the present study, $Zn_{1-x}Co_xO$ nanopowders with various Co-doping content (x = 0.07~0.15) were synthesized by sol-gel auto-combustion method. The effects of Co concentration on the phase and structure as well as the magnetic property of the as-synthesized powders were investigated by means of XRD, Raman, UV-vis-NIR, SEM and PPMS. It was found that Co-doping did not change the wurtzite structure of ZnO. Single-phase could be obtained for $Zn_{0.93}Co_{0.07}O$ and $Zn_{0.9}Co_{0.1}O$ nanopowders, but Co cluster began to appear when doped with more Co (x = 0.125, 0.15). The Co^{2+} substituted Zn^{2+} and was incorporated into the lattice of ZnO. The $Zn_{1-x}Co_xO$ nanopowders exhibited ferromagnetic behavior at room temperature, and the ferromagnetism is enhanced with increasing Co-doping content.

INTRODUCTION

Diluted magnetic semiconductors (DMSs)[1] are non-magnetic semiconductors doped with a few percent of magnetic elements, usually transition-metals (TM), which are expected to be not only easily integrable with existing semiconductors but also highly spin-polarized. Due to the unique semiconducting and ferromagnetic properties, they have attracted abundant studies in the past decade. Since the first discovery of room temperature ferromagnetism in Co:TiO_2 system[2], a considerable number of investigations on oxide-based DMSs have been triggered, such as TM-doped ZnO[3], SnO_2[4] and Cu_2O[5], etc.

ZnO-based DMSs doped with transition metals, in particular Co-doped ZnO, are of great interest since it has been predicted that they may be good ferromagnetic candidates with large magnetization and high Curie temperature over 300 K[6]. Co-doped ZnO diluted magnetic semiconductor powders can be prepared by physical or chemical techniques, including solid state reaction[7], hydrothermal method[8], coprecipitation technique[9] and sol-gel[10], etc. Complex schedules and low production rate are the common problems of those methods.

Sol-gel auto-combustion synthesis[11] method (also called low-temperature self-combustion or solution combustion) is a novel and simple way to quickly prepare good nanoparticles, where the chemical sol-gel and self-combustion reaction are combined. The advantages[12,13] of sol-gel auto-combustion include: good chemical homogeneity, high product purity and crystallinity, fine particle size and narrow particle size distribution, controllable stoichiometry, easily doping, simple equipment and preparation process, low processing time, low external energy consumption, etc.

During the last decade, the application of sol-gel auto-combustion method for the synthesis of nanosized powders has been used in an increasing intensity, but few reports on Co-doped ZnO nanopowders synthesized by this method are found. In the present study, Co-doped ZnO ($Zn_{1-x}Co_xO$) diluted magnetic semiconductor nanopowders were prepared by sol-gel auto-combustion synthesis method. The effects of Co concentration on the structure and magnetic properties of the $Zn_{1-x}Co_xO$ nanopowders were investigated.

EXPERIMENTAL PROCEDURES

Preparation

Co-doped ZnO diluted magnetic semiconductor nanopowders were prepared by sol-gel auto-combustion synthesis method. This method involves exothermic and self-sustaining thermally-induced anionic redox reaction of xerogel, which is obtained from aqueous solution containing desired metal salts (oxidizer) and organic complexant (reductant). Zinc nitrate and cobalt nitrate were used as the starting materials, which were dissolved in 2-methoxyethanol solution for the preparation of precursor sol. The Zn molar concentration in the solution was kept at 0.5 mol/L and the concentration of Co as the dopant was 7, 10, 12.5, 15 at %, respectively, with respect to the total amount of Zn and Co. The solution was stirred at 60 °C to yield a clear and homogeneous sol and further dried at 100 °C for 12 hours to evaporate water completely and form the gel, which subsequently underwent auto-combustion reaction to produce the fine $Zn_{1-x}Co_xO$ ($x = 0.07, 0.1, 0.125, 0.15$) nanopowders.

Characterization

The structure of the as-synthesized $Zn_{1-x}Co_xO$ ($x = 0.07\sim0.15$) nanopowders was analyzed by powder X-ray diffractometer with Cu Ka radiation (Rigaku D/Max-IIIA). Raman spectra were recorded using a Raman spectroscopy (Renishaw InVia) with an Ar^+ laser (wavelength 514.5 nm) as the excitation source. The morphology of the nanopowders was characterized by scanning electron microscopy (Hitachi S-3400). Optical absorption measurement was conducted at room temperature using a UV-visible-NIR spectrophotometer (Shimadzu UV-2550). Magnetic property was measured by a physical property measurement system (Quantum Design PPMS-9T).

RESULTS AND DISCUSSION

Fig.1 shows the XRD patterns of the $Zn_{1-x}Co_xO$ ($x = 0.07, 0.1, 0.125, 0.15$) nanopowders synthesized by sol-gel auto-combustion method. The diffraction patterns reveal the presence of the peaks corresponding to hexagonal ZnO and Co-doping does not change the wurtzite structure of ZnO. For $Zn_{0.93}Co_{0.07}O$ and $Zn_{0.9}Co_{0.1}O$ nanopowders, all the characteristic diffraction peaks are found to be in good agreement with the standard pattern of ZnO (JCPDS Card No. 89-0510), showing a single phase of ZnO. No second phase is detected in those samples, which suggests that Co^{2+} has been dissolved into the ZnO crystalline lattice. On the contrary, for the $Zn_{1-x}Co_xO$ nanopowders doped with more Co ($x = 0.125, 0.15$), the diffraction peaks attributed to Co cluster begin to appear and grow in intensity with the increasing of Co-doping content. The formation of Co may be resulted from the unreacted Co^{2+} at higher doping concentration, because the sol-gel auto-combustion process is likely to cause a hypoxic environment due to the violent reaction.

On the other hand, the lattice constant of *c*-axis (*d*(002) value) increases slightly with increasing Co concentration, implying that the ionic size of tetrahedrally coordinated Co^{2+} is larger than that of Zn^{2+} in the same symmetry[7].

Raman spectroscopy can be useful to assist the identification of crystal structure of Co-doped ZnO. ZnO has six Raman-active phonon modes in the first-order phonon, which are E_2(low), A_1(TO), E_1(TO), E_2(high), A_1(LO) and E_1(LO)[14]. Fig.2 shows the Raman spectra of $Zn_{0.93}Co_{0.07}O$ and $Zn_{0.9}Co_{0.1}O$ nanopowders, which are single-phased in wurtzite structure as illustrated by the XRD measurements (Fig.1). Three phonon signals located at 332, 437 and 500~600 cm^{-1} are observed, which can be assigned to E_2(high)-E_2(low), E_2(high), A_1(LO) and quasi-LO phonon modes[15], respectively.

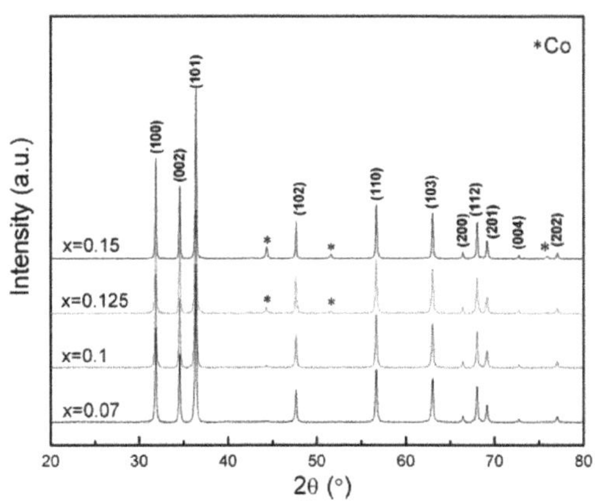

Fig.1 XRD patterns of $Zn_{1-x}Co_xO$ nanopowders

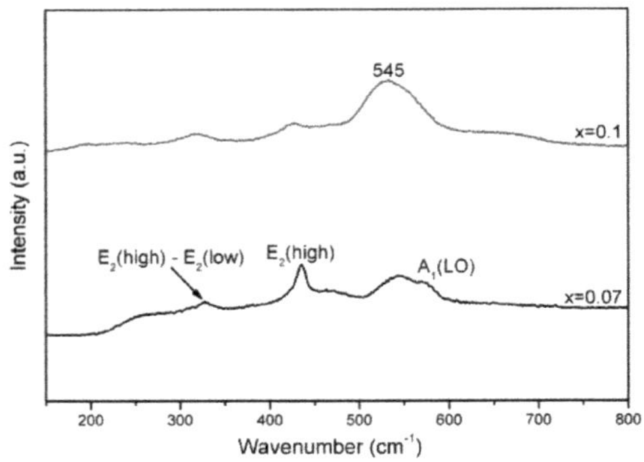

Fig.2 Raman spectra of $Zn_{1-x}Co_xO$ nanopowders

As seen from Fig.2, the E_2(high) phonon signal is broaden and weaken with the increasing of Co-doping content, which maybe resulted from the compositional fluctuation caused by the random substitution of Co^{2+} into the Zn^{2+} sites[16]. A broad signal at the wavenumber of 500~600 cm^{-1} can be resolved into two component bands: the A_1(LO) phonon mode near 575 cm^{-1}, and the quasi-LO phonon mode near 545 cm^{-1} due to the oxygen vacancies or zinc interstitials bounded on the tetrahedral Co sites. The intensities of both the A_1(LO) mode and quasi-LO mode are increased with increasing Co concentration, which gives a clear evidence for the Co substitution in ZnO host lattice[17]. In addition, there is no phonon signal corresponding to CoO or Co_3O_4[18], which is in consistence with the XRD results.

In order to further confirm the substitution of Co^{2+} in tetrahedral sites of the ZnO wurtzite structure, a UV-visible-NIR spectroscopy was used. Fig.3 shows the UV-vis absorption spectra of $Zn_{1-x}Co_xO$ (x = 0.07, 0.1, 0.125, 0.15) nanopowders. The absorption spectra exhibit three new absorption shoulders approximately at 566, 610 and 655 nm, which can be ascribed to $^4A_2(F)\rightarrow{}^2A_1(G)$, $^4A_2(F)\rightarrow{}^4T_1(P)$ and $^4A_2(F)\rightarrow{}^2E(G)$ transitions, respectively, for Co^{2+} with $3d^7$ high-spin configuration under the tetrahedral crystal field[19]. Additionally, the intensity of absorption peak increase with increasing Co content, which implies that more Co^{2+} substitutes Zn^{2+} in tetrahedral sites of the wurtzite structure.

Fig.4 shows the SEM morphology of $Zn_{0.93}Co_{0.07}O$ nanopowders prepared by sol-gel auto-combustion synthesis. The powders are irregular polygon and the particle size is uniform, which is about 60nm in average.

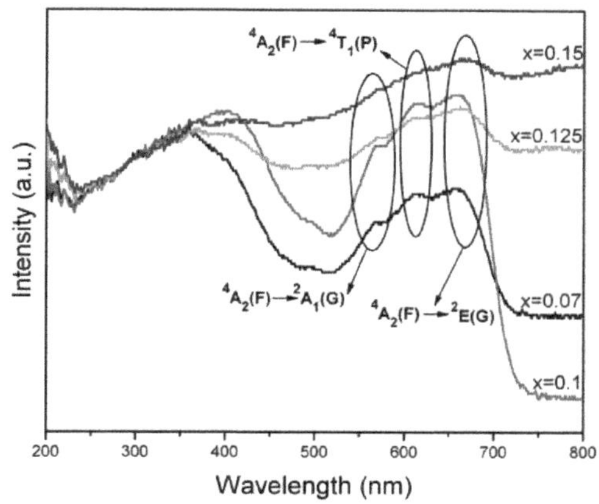

Fig.3 UV-vis absorption spectra of $Zn_{1-x}Co_xO$ nanopowders

Fig.4 SEM morphology of $Zn_{0.93}Co_{0.07}O$ nanopowders

The magnetic property of the as-synthesized $Zn_{1-x}Co_xO$ nanopowders was measured by PPMS at room temperature. Fig.5 shows the field dependence of specific magnetization (*M-H* curve) of two Co-doped compositions, i.e., $Zn_{0.93}Co_{0.07}O$ and $Zn_{0.9}Co_{0.1}O$. We can see that they show clear ferromagnetic *M-H* hysteresis in the field range of ±1600Oe. Outside this range, the specific magnetization increases with increasing magnetic field and shows the saturation magnetization of 0.065 and 0.138emu/g at 10kOe for $x = 0.07$ and 0.1, respectively. That is, the $Zn_{1-x}Co_xO$ ($x = 0.07$, 0.1) nanopowders exhibit ferromagnetic behavior at room temperature, and the magnetization is increased with increasing Co-doping concentration. Since there is no indication for the presence of Co metal or oxides in the $Zn_{1-x}Co_xO$ nanopowders according to the XRD patterns (Fig.1) and Raman spectra (Fig.2), it can be thus concluded that the ferromagnetism in the Co-doped ZnO should originate from the substitution of Co^{2+} into the Zn^{2+} sites. On the other hand, the $Zn_{0.9}Co_{0.1}O$ nanopowders synthesized in the present study by sol-gel auto-combustion method has a specific magnetization value of 0.124emu/g at 5kOe, which is much larger than that of 0.051emu/g[20] and 0.00256emu/g[21] reported in other $Zn_{1-x}Co_xO$ powders prepared by sol-gel method.

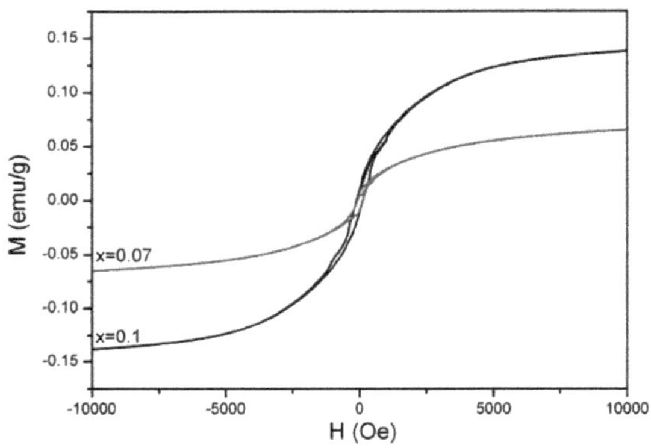

Fig.5 *M-H* curves of $Zn_{1-x}Co_xO$ nanopowders

CONCLUSIONS

We have prepared $Zn_{1-x}Co_xO$ (x = 0.07~0.15) diluted magnetic semiconductors nanopowders by sol-gel auto-combustion synthesis method and investigated the structure as well as magnetic property of the powders. Single-phased $Zn_{0.93}Co_{0.07}O$ and $Zn_{0.9}Co_{0.1}O$ nanopowders in wurtzite structure were obtained, but Co cluster began to appear at higher Co-doping content (x = 0.125, 0.15). In the as-synthesized $Zn_{1-x}Co_xO$ nanopowders, Co^{2+} had been dissolved into the ZnO crystalline lattice, and substituted the wurtzite structure Zn^{2+} sites into the tetrahedral coordination environment. The $Zn_{1-x}Co_xO$ (x = 0.07, 0.1) nanopowders exhibited ferromagnetic behavior at room temperature, and the ferromagnetism is stronger when doped with more Co.

ACKNOWLEDGEMENT

This work is financially supported by the Program for New Century Excellent Talents in University (NCET-10-0662), National Natural Science Foundation of China (51272195), the 111

Project (B13035) and the Fundamental Research Funds for the Central Universities (2012-Ia-012).

REFERENCES

[1]T. Dietl, A Ten-Year Perspective on Dilute Magnetic Semiconductors and Oxides, *Nat. Mater.*, **9**, 965-74 (2010).

[2]Y. J. Matsumoto, M. Murakami, T. J. Shono, Room temperature Ferromagnetism in Transparent Transition Metal-Doped Titanium Dioxide, *Science*, **291**, 854-6 (2001).

[3]K. Ueda, H. Tabata, T. Kawai, Magnetic and Electric Properties of Transition-Metal-Doped ZnO Films, *Appl. Phys. Lett.*, **79**, 988-90 (2001).

[4]S. B. Ogale, R. J. Choudhary, J. P. Buban, High Temperature Ferromagnetism with a Giant Magnetic Moment in Transparent Co-Doped $SnO_{2-\delta}$, *Phys. Rev. Lett.*, **91**, 077205-8 (2003).

[5]S. N. Kale, S. B. Ogale, S. R. Shinde, Magnetism in Cobalt-Doped Cu_2O Thin Flms without and with Al, V, or Zn Codopants, *Appl. Phys. Lett.*, **82**, 2100-2 (2003).

[6]T. Dietl, H. Ohno, F. Matsukura, J. Cibert, D. Ferrand, Zener Model Description of Ferromagnetism in Zinc-Blende Magnetic Semiconductors, *Science*, **287**, 1019-22 (2000).

[7]S. J. Luo, C. B. Wang, X. Zhou, Q. Shen, L. M. Zhang, Fabrication and Characterizations of $Zn_{1-x}Co_xO$ Bulk Ceramics Prepared by Solid State Reaction Combined with Spark Plasma Ssintering, *J. Mater. Sci.: Mater. Electron.*, **23**, 1477-84 (2012).

[8]H. Q. Li, H. Y. Miao, G. Q. Tan, Preparation of $Zn_{1-x}Co_xO$ Diluted Magnetic Semiconductor by Hydrothermal Method, *Key Eng. Mater.*, **368-372**, 620-2 (2008).

[9]M. Bouloudenine, N. Viart, S. Colis, J. Kortus, A. Dinia, Antiferromagnetism in Bulk $Zn_{1-x}Co_xO$ Magnetic Semiconductors Prepared by the Coprecipitation Technique, *Appl. Phys. Lett.*, **87**, 052501-3 (2005).

[10]O. D. Jayakumar, S. N. Achary, C. Sudakar, Experimental and Theoretical Investigations on Magnetic Behavior of (Al,Co) Co-Doped ZnO Nanoparticles, *Nanoscale*, **2**, 1505-11 (2010).

[11]S. T. Aruna, A. S. Mukasyan, Combustion Synthesis and Nanomaterials, *Curr. Opin. Solid St. M.*, **12**, 44-50 (2008).

[12]A. S. Mukasyan, P. Epstein, P. Dinka, Solution Combustion Synthesis of Nanomaterials, *P. Combust. Inst.*, **31**, 1789-95 (2007).

[13]A. Sutka, G. Mezinskis, A. Pludons, Characterization of Sol-Gel Auto-Combustion Derived Spinel Ferrite Nano-Materials, *Power Eng.*, **56**, 254-9 (2010).

[14]T.C. Damen, S.P.S. Porto, B. Tell, Raman Effect in Zinc Oxide, *Phys. Rev.*, **142**, 570-4 (1966).

[15]J.M. Calleja, M. Cardona, Resonant Raman Scattering in ZnO, *Phys. Rev. B*, **16**, 3753-61 (1977).

[16]N. Hasuike, K. Nishio, H. Katoh, Structural and Electronic Properties of ZnO Polycrystals Doped with Co, *J. Phys. Condens. Matter.*, **21**, 064215-9 (2009).

[17]K. Samanta, P. Bhattacharya, R.S. Katiyar, W. Iwamoto, P.G. Pagliuso, C. Rettori, Raman

Scattering Studies in Dilute Magnetic Semiconductor $Zn_{1-x}Co_xO$, *Phys. Rev. B*, **73**, 245213-5 (2006).
[18]C.W. Tang, C.B. Wang, S.H. Chien, Characterization of Cobalt Oxides Studied by FT-IR, Raman, TPR and TG-MS, *Thermochim. Acta*, **473**, 68-73 (2008).
[19]P. Koidl, Optical Absorption of Co^{2+} in ZnO, *Phys. Rev. B*, **15**, 2493-9 (1977).
[20]S. Maensiri, P. Laokul, S. Phokha, A Simple Synthesis and Magnetic Behavior of Nanocrystalline $Zn_{0.9}Co_{0.1}O$ Powders by Using Zn and Co Acetates and Polyvinyl Pyrrolidone as Precursors, *J. Magn. Magn. Mater.*, **305**, 381-7 (2006).
[21]D. A. Schwartz, N. S. Norberg, Q. P. Nguyen, Magnetic Quantum Dots: Synthesis, Spectroscopy, and Magnetism of Co^{2+}- and Ni^{2+}-Doped ZnO Nanocrystals, *J. Am. Chem. Soc.*, **125,** 13205-18 (2003).

Synthesis and Processing of Materials Using Electric Fields/Currents

ADVANCED USAGE OF SPS TECHNOLOGY FOR PRODUCING INNOVATIVE MATERIALS

Foad NAIMI, Ludivine MINIER, Cédric MORIN, Sophie LE GALLET and Frédéric BERNARD

Laboratoire Interdisciplinaire Carnot de Bourgogne, UMR 6303 CNRS / Université de Bourgogne – 9 Avenue A. Savary BP 47870 21078 Dijon cedex

ABSTRACT

It is undeniable that the usage of SPS authorizes exceptional performances such as the strong limitation of the grain growth, the fabrication of functional graded materials, the synthesis of various innovative materials and, at last, the possibility to carry out homogenous or heterogeneous joining without matter provision. Nevertheless, the phenomena that govern the sintering mechanisms are still poorly known and are the object of numerous researches in order to control this process that allows sintering, assembling, and synthesizing different types of materials (metals, ceramics, and composites). Many examples will come to illustrate these applications, obtained by the technology SPS available in Dijon, which allows submitting the sample to a DC pulsed current (0-8 V, 1-24 kA) while applying a uniaxial load (16 to 1250 kN) and, this in a range of varying temperature from the ambient one until 2 400 °C. The sintered samples can have a diameter varying from 30 to 150 mm. The experiences generally are performed under vacuum but it is possible to work under controlled atmosphere.

INTRODUCTION

Spark Plasma Sintering (SPS) is one of the most attractive techniques for producing innovative materials with a moderate cost and, for some cases, enhanced properties. This fast sintering technique is a high-speed powder consolidation/sintering technology capable of processing conductive and nonconductive materials. SPS consists in applying simultaneously a pulsed direct current (DC) of high intensity (several thousands of amperes) and a uniaxial pressure to the powder (Figure 1).

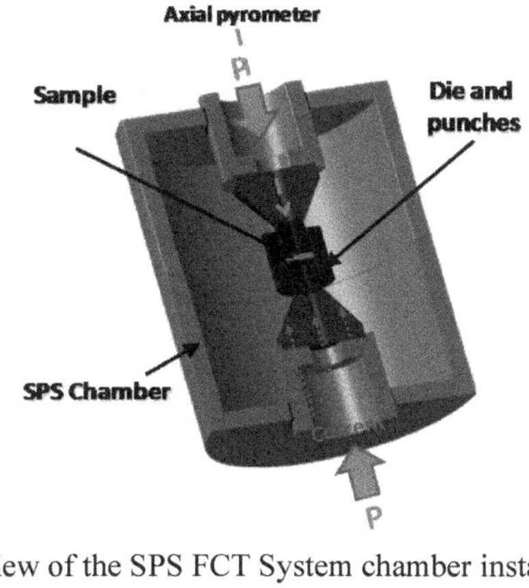

Figure 1: View of the SPS FCT System chamber installed in Dijon

This very fast process allows the fabrication of materials such as nanostructured materials, functionally graded materials... Moreover, SPS can potentially produce near-net and net shapes directly from powders[1,2]. The powder is loaded in a cylindrical die (usually graphite). The upper and lower punches through which the pressure (typically 1-200 MPa) is applied, also act as electrodes which bring the electric current in the die and sample (Figure 1). The Joule effect enables very quick heating of the system (typically up to 500 °C/min). SPS experiments are generally performed under vacuum or under controlled atmospheres. The electric current is applied as high current (~1000 to 30 000 A) – low voltage (0 – 10 V) DC short pulses, which are expected to lead to "sparks" and "plasmas" (the effects from which derives the name spark plasma sintering of the technique). The size of the samples is 10-50 mm in research devices; it can reach more than 300 mm in industrial equipments.

Although the first patents on "Electric current activated/assisted sintering (ECAS)" have been published in 1906 [3], SPS has introduced an alternative route in the synthesis of new or innovative materials in the last 10 years only. Its extreme rapidity and the possibility to sinter a large class of materials (metals, ceramics, even polymers) have driven a rapid emergence of this technique all around the world, especially, in South-East Asia during the last part of 20th century [1,4]. In a recent review containing 1005 papers, Orru et al.[4], report the capability of this technology to produce "in a short time" various types of materials.

Concerning the understanding of this fast sintering method, this paper reviews studies on the influence of the operating conditions such as pressure, temperature, heating rate, current intensity, pulses… and, on modelling approaches by finite element methods. However, this paper points also on the lack of knowledge of the physical mechanisms involved during the SPS process. Sintering is often described by 3 stages: i) neck formation/growth at the contact points between initial particles, ii) densification (pore elimination) and iii) grain coarsening. Necks are places where the inter-particle boundaries are formed, which ensure the powder cohesion. They also play a crucial role of channels for mass transfer between particles. For these reasons, neck formation is often considered as the first and key step of sintering. Densification subsequently occurs through several mechanisms, under various driving forces, many of them being also driving forces for grain coarsening (Figure 2).

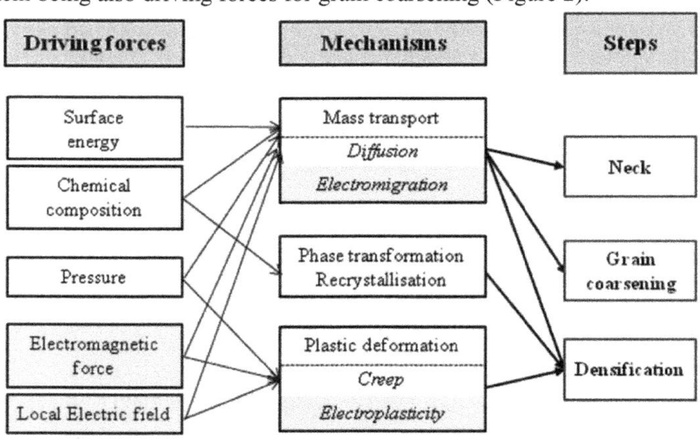

Figure 2: Simplified description of driving forces and sintering mechanisms, and their effect on sintering steps. The yellow background marks driving forces and mechanisms which are specific to SPS [6].

The common driving force to all sintering techniques is the surface energy, related to geometrical parameters: local curvature, grain size... They induce numerous mass transport mechanisms (surface, bulk and grain boundary diffusion), which contribute to sintering phenomena. In the case of multi-element systems, composition gradients are supplemental driving forces for mass transport and phase transformations. In pressure assisted sintering techniques, such as hot pressing and SPS, the applied pressure is an additional driving force for mass transport and plastic deformation during densification. SPS brings two specific driving forces, related to the electric current inside the sample: i) constrictive macroscopic electromagnetic forces [5] which act in a similar way as assisting pressure, ii) local electric field and current which induce electromigration mechanisms and electroplasticity.

The efficiency of the SPS technique is however widely attributed to other effects, which are not specific mechanisms, but enhancing phenomena. The "spark and plasma" formation induced by local electromagnetic fields between particles is expected to induce cleaning and melting effects, which enhance first neck formation and then almost all sintering mechanisms. Even in the absence of spark and plasma, the concentration of high currents in small particles or in necks leads to local overheating and even melting of key areas which favour diffusion as well as plastic deformation. Last but probably not the least, the very high heating rates (up to several hundreds of K per min) induce thermal gradients which contribute to mass transport and favour the densification mechanisms with respect to particle growth ones [7].The most common SPS mechanisms pointed out by the literature can be classified in the three following groups.

(i) Mass transport mechanisms. The electric current has an overall enhancing effect on atom and defect mobility, through the high temperatures locally reached by Joule effect [2,8]. Current-induced mass transport (electro-migration effect), a well-known problem in electronics leading to failure of Cu interconnects [9-10], is a potential mechanism of mass transport specifically involved in SPS. It has been observed macroscopically by dedicated electromigration experiments in many metallic systems: Sn-Cu and Sn-Ni [11], Sn-Ag [12], Al-Au [13], Ni-Ti [14], Mo-Si [15] and Cu-Ni [16].

(ii) Plasticity, recrystallization, phase transformations. Classical mechanisms operating at high temperature and high stress, like diffusion and dislocation creep, dynamic recrystallisation and phase transformation, can also be enhanced by the electric current. The electroplasticity effect facilitates the deformation of powder particles and enhances the densification step, by lowering the mechanical strength of the material [17-21]. As illustrated by Jabber et al. [22], strongly deformed areas and recrystallized grains in TiAl alloys sintered by SPS have been clearly evidenced. The observation of the periphery of powder particles and necks shows an accumulation of zones with different phases, due to phase transformations, chemical segregations… Grosdidier et al. [23] and Song et al. [24] also observed the formation of fine grains, localised within neck zones.

(iii) Melting and plasma formation. Electric sparks and plasmas around the powder particles are assumed to clean the powder surface, to destroy oxide layers [25-26] or to induce local melting of the particle surface, which could accelerate densification [27]. For submicrometric and nanometric powders, the breakdown voltage could be reached, leading to an electrical discharge inside the powder and to plasma formation, even for low applied voltages [28]. However, these effects are subject of controversy and, have not been supported by any compelling evidence [2,4,21]. Using dedicated in situ atomic emission spectroscopy and ultrafast in situ voltage measurements, Hulbert et al. [29] concluded to the absence of plasma during SPS. To our knowledge, no careful microstructure analysis (e.g. by high resolution electron microscopy on necks), aiming at a systematic search for plasma or melting evidences, has ever been carried out. It must also be noted that, even without plasma and spark, local heating at necks can lead to local melting.

Even if many works are still necessary to understand the role of the current on sintering stages, SPS technology opens original routes in materials science. Especially, in this paper, some interesting examples will be described in the case of the production of nanostructured materials, the synthesis-consolidation in one step and, finally for the joining of homogeneous and heterogeneous systems.

SINTERING OF DENSE NANOSTRUCTURED NICKEL

One of the challenges for many groups is to control the grain size of materials during the fabrication process in order to obtain dense nanostructured materials having improved mechanical properties. An alternative method for producing dense nanostructured materials is a mechanically activated spark plasma sintering process[30]. Such a process starting from micrometric commercial powders consists in two main steps: (a) a mechanical activation (MA) using a high energy ball mill and (b) a densification using the SPS device.

Conditions of powder preparation were selected to produce per vial one batch necessary to perform one SPS sample. The dimensions of sintered specimens must be sufficient to perform three tensile tests i.e. 50 mm in diameter and 10 mm in height. Consequently, 120 g of elemental powder Ni (Alfa Aesar, APS 3-7µm in particle size, 99.9% purity) was milled in a planetary ball vario-mill Fritsch Pulverisette 4 [31]. Based on previous works [32], a specific ball-milling condition was established at 250 rpm (rotation per minute) for the disk rotation speed and -50 rpm for the absolute vial rotation speed. The charge ratio C_R (steel ball to powder mass ratio) was 7. The powder being constituted of quite abrasive materials, a short duration of milling has been selected to avoid any contamination of the product by the milling tools (8 h uninterrupted). Mechanically activated powders in comparison to the commercial powder appear as large agglomerates due to successive fracture / welding process occurring during the milling (Figure 3).

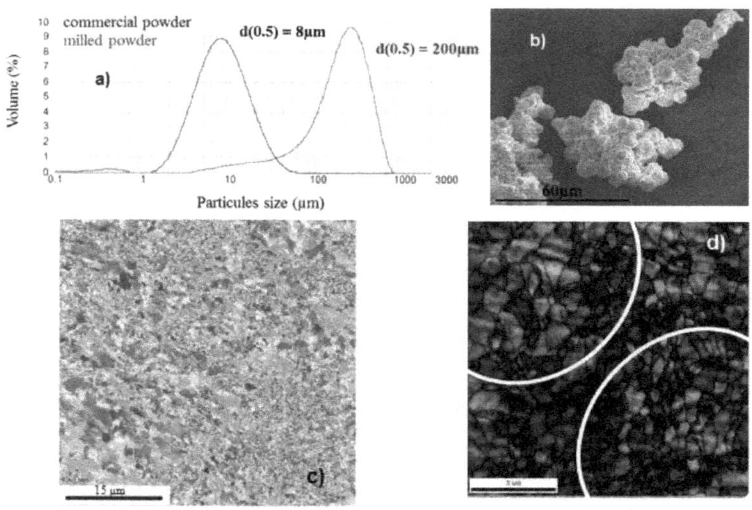

Figure 3 : a) Particle size distribution determined by laser granulometry for both systems, b) SEM observations of as-milled Ni powder, c) EBSD observation (inverse pole figure mode) of mechanically activated Ni agglomerate and d) EBSD-SEM images of the dense sample sintered from the milled powder under 200MPa in which one zone composed of large grains and another zone with smaller grain were circled.

Indeed, the size of commercial powder grain increases from 8 μm to 200 μm whereas the crystallite size determined from XRD peak profile analysis decreases from 250 nm to 52 nm. Then, the powders were first cold compacted in the graphite die at the sintering pressure during 5 minutes. The temperature was monitored by a K-type thermocouple located in a hole at 3 mm far from the sample and half height of the die. The optimization of the sintering conditions of the Ni powder has been performed in a previous work by the establishment of a sintering map [33-34]. Accordingly, the as-milled Ni powder was sintered at 750°C under 200MPa during 5 minutes. The heating rate was divided in two steps. The heating rate was 200°C/min until 600°C and decreased at 50°C/min for the second step. The density of the sample was 98 ± 0.5 %. These latter presented a specific and interesting microstructure which combines larger particles with smaller ones as be shown on the SEM-EBSD image presented Figure 3d.

Finally, these Ni-nanostructured samples were tested using classical static mechanical tests such as traction tests and compared to literature data. Tensile tests were performed at a strain rate of 10^{-3} s^{-1} with specimens 16 mm in gage length and 4 mm in gage diameter using a Testwell machine equipped with a load cell of capacity 5000 daN. The gage length elongation was measured with an extensometer Epsilon over a gage length of 10 mm. These nanostructured nickel specimens which had yield strength of 600 MPa with ductility close to 30% are in agreement with the expected Hall-Petch relationship (Figure 4).

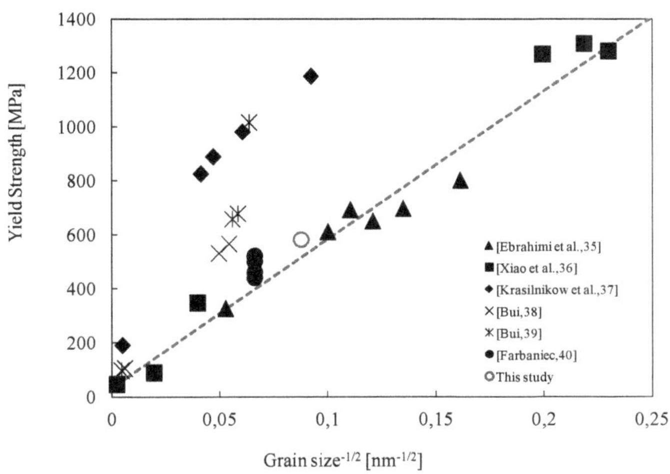

Figure 4 : Hall-Petch plot for nickel processed by different processing routes [35-40].

REACTIVE SINTERING FOR CONDITIONING IODINE IN AN APATITE LIKE CERAMIC

Since 2005, we collaborate with the laboratory of nuclear waste conditioning at the CEA Marcoule on the elaboration by FAST of ceramic matrix for the immobilization of iodine (one of the envisaged strategies for the management of the long life nuclear wastes and their storage in deep geological layers). The safety of storage rests on a packaging in a chemically inert matrix and lacking in open porosity. Nevertheless, the densification of iodine based materials requires low sintering temperatures because the decomposition of the iodoapatite ($Pb_{10}(VO_4)_{4.8}(PO_4)_{1.2}I_2$) under air begins as early as 500 °C. Previously, it was investigated to consolidate this iodoapatite only in confined environment. The confinement of the environment can be performed in lead phosphovanadate ($Pb_3(VO_4)_{1.6}(PO_4)_{0.4}$), that itself consolidates at low temperature. This is for the reason that a SPS parametric study of lead phosphovanadate milled powders, used as material of gangue, was carried out. The milling conditions of lead phosphovanadate were adapted in order to reach a sintering temperature compatible with thermal stability of the iodoapatite. Samples core+gangue were thus sintered by SPS with success [41-42]. A fine description of shrinkage curve of this material has shown that it was possible to dissociate the behavior of the gangue from that of the core (Figure 5).

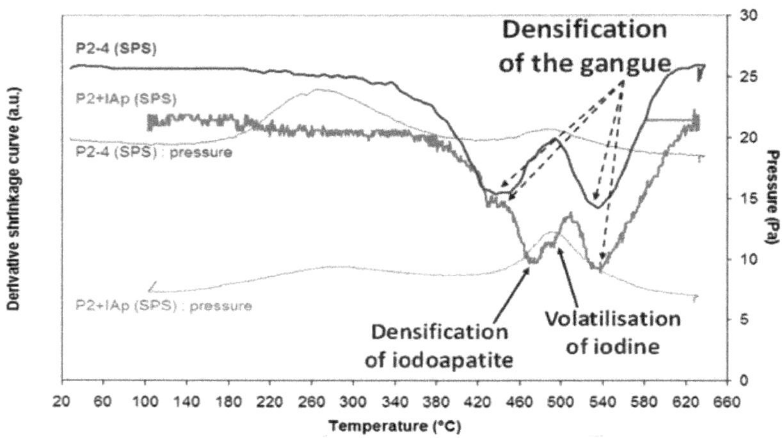

Figure 5 : Shrinkage derivatives and pressure variations in the furnace versus temperature for gangue material alone (P2-4) and for gangue+core (P2 + IAp).

It appeared that it was theoretically possible to realize the sintering of the iodoapatite in non-confined environment at lower temperatures to iodoapatite decomposition during the SPS sintering. However, full densification could not be obtained from direct sintering of a iodoapatite powder (5.8 $m^2 \cdot g^{-1}$ (BET(N_2)). Iodoapatite was successfully consolidated by reactive spark plasma sintering only [42-43]. The initial powder consisted of a mixture of 3 moles of $Pb_3(VO_4)_{1.6}(PO_4)_{0.4}$ per mole of PbI_2, corresponding to the stoichiometry of equation (1).

$$3Pb_3(VO_4)_{1,6}(PO_4)_{0,4} + PbI_2 \rightarrow Pb_{10}(VO_4)_{4,8}(PO_4)_{1,2}I_2 \qquad \textbf{(1)}$$

Further details on the synthesis of the reagents can be found in reference [43]. Different attrition milling conditions were used to obtain two initial specific surface areas: 3.3 $m^2 \cdot g^{-1}$ and 9.5 $m^2 \cdot g^{-1}$ (BET(N_2)). For each test, 1.675 grams of powder were sintered under vacuum in graphite molds (internal diameter of 10mm) with a SPS 515S machine (Syntex Inc.). The powder layer was first compacted at room temperature before each experiment. Densities of around 97% were obtained for both specific surface areas. The ceramic materials obtained under these conditions (pellets 10 mm in diameter and 3 mm thick) were of uniform composition (single-phase) and microstructure (120 ± 40 nm) as observed on the Figure 6.

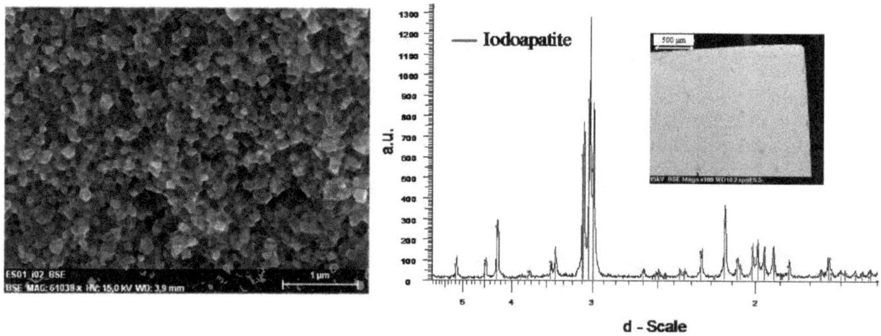

Figure 6 : SEM images and XRD pattern of dense $Pb_{10}(VO_4)_{4.8}(PO_4)_{1.2}I_2$ prepared by reactive sintering. A dense iodoapatite of 97% with a fine and homogeneous microstructure have been obtained [43].

Despite a complex sintering process involving multiple steps described below, the production of a given microstructure seems not to depend on the powder characteristics (initial specific surface area). The first step is the coalescence of lead iodide at temperatures below 250 °C, followed by the iodoapatite synthesis reaction between 250 and 390 °C together with initial sintering in the solid phase up to about 350 °C, and then the appearance of a transient liquid phase above 350 °C that accelerates the reaction and densification kinetics. The sintering temperatures ranging from 350 to 450 °C (depending on the other sintering parameters) for both powders, the achievement of the densification occurs systematically in presence of a liquid phase, which seems to erase the initial state of the powder.

HOMOGENEOUS AND HETEROGENOUS JOINING BY SPS

Recently, our research team was also very interested in the capability offered by the SPS technique to associate different kinds of materials that can present themselves in the form of a powder or massive just as well in the domain of the ceramics as that of the metals. In fact, through the technology SPS, different homogenous or heterogeneous assemblies can be imagined. Let's quote for example the works in the framework of the program of research "new generation of nuclear reactors", for which the assembly of two plates of SiC was evaluated. This study clearly showed that it is possible to eliminate the interface between these two SiC-ceramics while playing on the parameters (i.e. temperature and pressure) as shown on the figure 7.

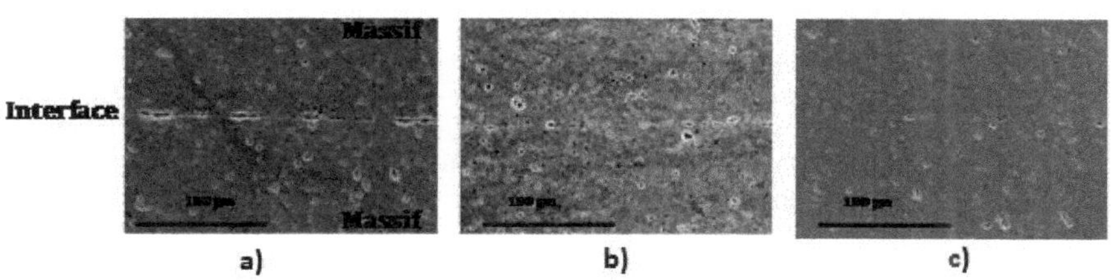

Figure 7 : SEM images of the interface between two SiC plates joined by SPS changing only the temperature b) and simultaneously the temperature and the uniaxial pressure c) in comparison with conditions established in a).

Usually, other kinds of assemblies are currently evaluated as the heterogeneous assembly between two different metallic alloys, copper alloys and titanium alloys respectively (Figure 8). There again while playing on the parameters of the SPS (temperature, pressure, time) but also on the surface state, for example roughness and surface cleanliness, it is possible to limit the formation of intermetallics that are prejudicial to interface resistance under mechanical solicitations. In addition, the other interest of this technology is the possibility to perform the assembly when the interface between these two alloys is perpendicular but also parallel to the applied force and to the current lines.

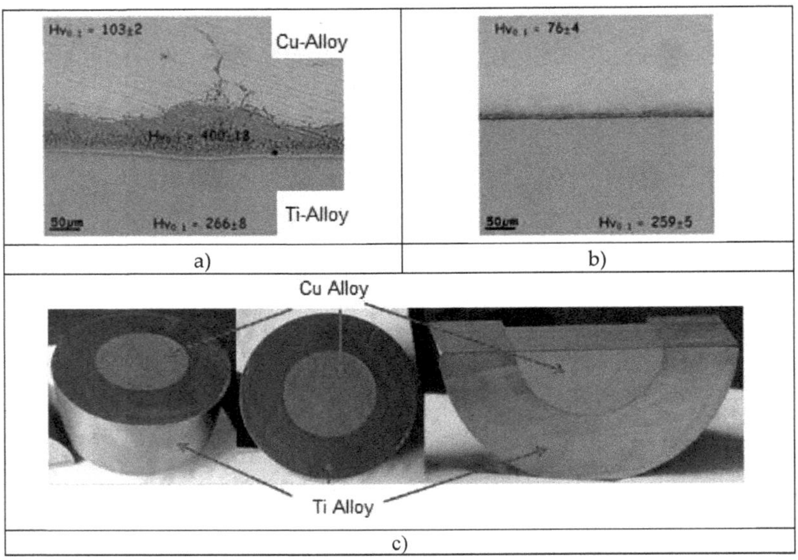

Figure 8 : Examples of heterogeneous joining obtained by SPS between two different metallic alloys with an interface perpendicular to the applied load (a). The surface preparation (chemical cleaning, mechanical polishing) can limit the intermetallic formation and/or can decrease the interface thickness (b) Joining parallel to the applied load (c).

SINTERING OF ALUMINA BASED FGM

A Functionally Graded Material (FGMs) is a material with a gradient of composition, phase, porosity or texture which leads to a gradient of properties like hardness, density, thermal conductivity, elasticity ... [44]. The integration of materials with different functional properties within the same product is a growing area in the manufacturing industry as nuclear, defense or aeronautic. By the control of the preparation of FGM, the quantity of raw material can be reduced and the functionality of components can be improved while optimizing the geometry and the distribution of the different materials from the design phase to the end-product. FGMs offer the advantage to save raw materials and to optimize the functionality by a suitable distribution of the materials within a product. The functional gradient may result from a microstructure gradient (grain size, porosity,) but it most often originates from a chemical composition gradient. According to the processing of materials, the composition may vary continuously or in stages. In a recent article, Kieback [45] classifies FGM in two categories, according to the type of gradients: (i) a gradient of chemical composition in a single-phase material, when a solid solution between the chemical constituents exists in the composition range; (ii) a gradient of the volume fractions of different phases in a multi-phase material.

According to Minier et al. work [34], the commercial alumina powder (Baikowski – BMA 15, 16.5 m²/g, D(50)=150 nm) was sintered at 1250°C under 65MPa during 10 minutes. The heating rate was 50°C/min until 1250°C. In order to manage temperature distribution on FGM within die and sample, a specific SPS die was designed in agreement with works published by Tokita [1].

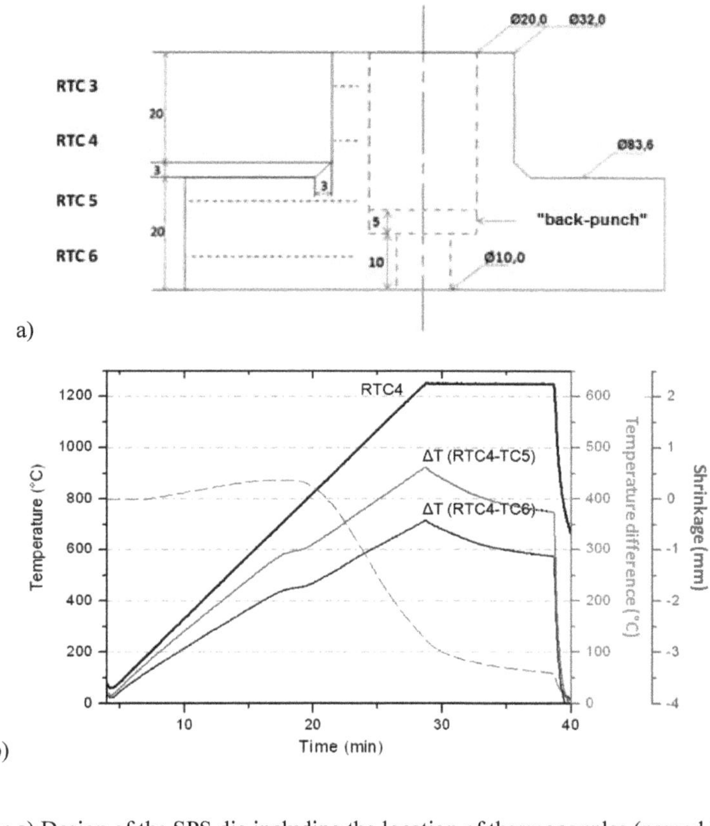

Figure 9 : a) Design of the SPS die including the location of thermocouples (named RTCX) and b) SPS cycle, thermal gradient inside the die and shrinkage of the sample

In order to prevent a heating source from the bottom of the system punch/sample/die the "back punch" die has been designed with an integrated puncher (Figure 9a). Figure 9b gives the temperature of the regulation thermocouple (RTC4) and the temperature difference between the regulation temperature and positions 5 and 6 during the SPS cycle. The shrinkage has been also reported on this figure. We can see that the densification of the alumina powder is observed when RTC4 reaches 700°C. Moreover, an accident can be observed in this linear procession of the difference temperature curves. Nevertheless, this temperature does not correspond with the temperature of sintering such a powder (880 °C under 70 MPa [46]). However, the green sample thickness is about 12 mm and 2 mm are in the high area above RTC 4 where the temperature is higher than 700 °C, which could been explained that the sintering begins. The temperature differences inside the die increase linearly with the 50 °C/min heating rate and the thermal difference ΔT RTC4-TC5 and ΔT RTC4-TC6 inside the die decrease along the 1250 °C stage. Actually, a high heating rate permits to obtain a maximal thermal gradient by limiting the thermal conduction contribution [8].

Figure 10 Microstructure along the sample sintered in the "back puncher" die.

This observation allows showing that the thermal gradient presents inside the die corresponds to a thermal gradient inside the sample. The microstructure can be directly related to the temperature difference inside the system. Near bottom puncher the grain size stay very small (200 nm) and they grow up with the temperature in the same time the sintering occur until 1.3 μm. The grains in contact with the upper puncher seem to be subject to a decrease of the temperature. This could be due to the presence of the puncher which decreases the current density against the small wall die. The sintered sample resulting of this SPS cycle has a relative density of 91 %. Now, further experiments and complementary simulations will be engaged in order to improve this concept.

CONCLUSION

Moreover, other promising programs dedicated to the fabrication by an electrical field assisted sintering process of advanced materials such as nanomaterials, composites and FGM, are engaged everywhere. The integration of advanced materials with improved properties is a growing area in the manufacturing industry as nuclear, defense or aeronautics. By the control of the preparation of these materials, the quantity of raw material can be reduced and the functionality of components can be improved while optimizing the geometry and the distribution of the different materials from the design phase to the end-product. Furthermore, the one-step manufacturing using powder metallurgy technique is a promising mean to generate homogeneous materials in an economically advantageous way. Besides, for thermal and structural applications, a "smooth and mastered" transition between the different materials can then be obtained by the introduction of a compositional gradient that limits thermal expansion differences and reduces thermal stresses during various solicitations.

REFERENCES

[1]M. Tokita, Trends in advanced SPS spark Plasma Sintering Systems and Technology - Functionally gradients Materials and Unique Synthetic Processing methods from next generation of powder technology, *Jl of the Soc. of Powder Technology*, Japan, **30**, 790 (1993).
[2]Z.A. Munir, U. Anselmi-Tamburini and M. Ohyanagi, The effect of electric field and pressure on the synthesis and consolidation of materials: A review of the spark plasma sintering method, *J. Mater. Sci.* **41**, 763-777 (2006).
[3]S. Grasso, Y. Sakka and G. Maizza, Electric current activated/assisted sintering (ECAS): a review of patents 1906–2008, *Science and Technology of Advanced Materials*, **vol. 10, no. 5**, Article ID 053001 (2009).
[4]R. Orru, R. Licheri, A. Mario Locci, A. Cincotti and G. Cao, Consolidation/synthesis of materials by electric current activated/assisted sintering, *Mat. Sci. Eng.*, **R63**, 127-287 (2009).
[5]S. Clyens, S.T.S. Al-Hasani and W. Johnson, The compaction of powder metallurgy bars using high voltage electrical discharges, *Int. J. Mech. Sci.* **18**, 37-40 (1976).
[6] Mechanisms of Spark Plasma Sintering in metallic materials, MF2 ANR-2011-BS09-020-02 program
[7]E.A. Olevsky, S. Kandukuri and L. Froyen, Consolidation enhancement in spark-plasma sintering: Impact of high heating rates, *J. Appl. Phys.* **102**, 114913 (2007).
[8] U Anselmi-Tamburini, S Gennarib, J E Garay and Z A Munir, Fundamental investigations on the spark plasma sintering/synthesis process II. Modeling of current and temperature distributions, *Mater. Sci. and Eng.* **A394**, 139-148 (2005).
[9]R. Frankovic and G. Bernstein, Temperature Dependence of ... Silicon Bipolar Technology, *IEEE Trans. Electron Devices*, **43(9)**, 1533-1538 (1996).
[10]K.C. Chen, C.N. Liao, W.W. Wu and L.J. Chen, Direct observation of electromigration-induced surface atomic steps in Cu lines by in situ transmission electron microscopy, *App. Phys. Lett.*, **90**, 1-3 (2007).
[11]S.W. Chen, C.M. Chen and W.C. Liu, Electric current effects upon the Sn/Cu and Sn/Ni interfacial reactions, *J. Elec. Mater.*, **27**, 1193-1199 (1998).
[12]C.M. Chen and S.W. Chen, Electric current effects on Sn/Ag interfacial reactions, *J. Elec. Mater.*, **28**, 902-906 (1999).
[13]N. Bertolino, J. Garay, U. Anselmi-Tamburini and Z.A. Munir, Electromigration effects in Al-Au multilayers, *Scripta Mater.*, **44**, 737-742 (2001).
[14]J. E Garay, U. Anselmi-Tamburini and Z.A. Munir, Enhanced growth of intermetallic phases in the Ni–Ti system by current effects, *Acta Materialia*, **51**, 4487-4495 (2003).
[15]W. Chen, U. Anselmi-Tamburini, J.E. Garay., J.R. Groza and Z.A. Munir, Fundamental investigations on the spark plasma sintering/synthesis process I. Effect of dc pulsing on reactivity, *Materials Science and Engineering*, **A394**, 132–138 (2005).
[16]J. Zhao, JE. Garay, U. Anselmi-Tamburini and Z.A.Munir, Directional electromigration-enhanced interdiffususion in the Cu-Ni system, *J. Appl. Phys.*, **102**, 114902-1-7 (2007).
[17]H. Conrad, A.F. Sprecher, W.D. Cao and X.P. Lu, Electroplasticity—the effect of electricity on the mechanical properties of metals, *JOM,* **42**, 28-33 (1990).
[18]B.R. Livesay, N.E. Donlin, A.K. Garrison, H.M. Harris and J.L. Hubbard, *International Reliability Physics Symposium, IEEE* **217**, (1992).
[19]H. Conrad, Effect of electric current on solid-state phase transformations in metals, *Mater. Sci. Eng.*, **A287**, 227–237 (2000).
[20]G. Tang, J. Zhang, Y. Yan, H. Zhou and W. Fang, The engineering application of the electroplastic effect in the cold-drawing of stainless steel wire, *J. Mat. Process. Tech.,* **137**, 96-99 (2003).

[21]M. Eriksson, Z. Shen and M. Nygren, Fast densification and deformation of titanium powder, *Powder Metallurgy*, **48**, 231-236 (2005).
[22]H. Jabbar, J.P. Monchoux, F. Houdellier, M. Dollé, FP. Schimansky, F. Pyczak, M. Thomas and A. Couret, Microstructure and mechanical properties of high niobium containing TiAl alloys elaborated by spark plasma sintering, *Intermetallics*, **18**, 2312-2321, (2010).
[23]T. Grosdidier, G. Ji, and S. Launois, Processing dense hetero-nanostructured metallic materials by spark plasma sintering, *Scripta Materialia*, **57**, 525–528 (2007).
[24]X. Song, X. Liu and J. Zhang, Neck Formation and Self-Adjusting Mechanism of Neck Growth of Conducting Powders in Spark Plasma Sintering, *J. Am Ceram. Soc.*, **89**, 494-500 (2006).
[25]M. Omori, Sintering, consolidation, reaction and crystal growth by the spark plasma system (SPS), *Mater. Sci. Eng.*, **A287**, 183-188 (2000).
[26]J.R. Groza and A. Zavaliangos, Sintering activation by external electrical field, *Mater. Sci. Eng.*, **A287**, 171-177 (2000).
[27]T. Grosdidier, G. Ji, F. Bernard, E. Gaffet, Z.A. Munir and S. Launois, Synthesis of Bulk FeAl nanostructured materials by HVOF spray forming and spark plasma sintering, *Intermetallics*, **14**, 1208-1213 (2006).
[28]R. Chaim, Densification mechanisms in spark plasma sintering of nanocrystalline ceramics, *Mat. Sci. Eng.*, **A443**, 25-32 (2007).
[29]D. M. Hulbert, A. Anders, J. Andersson, E. J. Lavernia, and A. K. Mukherjee, A discussion on the absence of plasma in spark plasma sintering, *Scripta Materialia*, **60**, 835–838 (2009).
[30]E. Gaffet, F. Bernard, J.C. Niepce, F. Charlot, Ch. Gras, G. Le Caer, J.L. Guichard, P. Delcroix, A. Mocellin andO. Tillement, some recent developments in mechanical activation and mecanosynthesis, *J of Mater. Chem.*, **9**, 305-314 (1999).
[31]M. Abdellaoui and E. Gaffet, The physics of mechanical alloying in a planetary ball mill: Mathematical treatment, *Acta Materiala*, **43**, 1087-1098 (1995).
[32]L. Minier, Influence du frittage « Flash » sur l'obtention de nanostructures dans des systèmes métalliques et céramiques, *Ph-D Thesis*, Université de Bourgogne, (2008).
[33]L. Minier, S. Le Gallet, Yu. Grin and F.Bernard, Influence of the current flow on the SPS sintering of a Ni powder, *J. of All. Comp.*, **508**, 412-418 (2010).
[34]L. Minier, S. Le Gallet, Yu. Grin and F.Bernard, A comparative study of nickel and alumina sintering using Spark Plasma Sintering (SPS), *Mater. Chem. Physics*, **134**, 243-253 (2012).
[35]F. Ebrahimi, G.R. Bourne, M.S. Kelly and T.E. Matthews, Mechanical properties of nanocrystalline nickel produced by electrodeposition, *Nanostructured Materials*, **11**, 343–350 (1999).
[36]C. Xiao, R.A. Mirshams, S.H. Whang and W.M. Yi, Tensile behavior and fracture in nickel and carbon doped nanocrystalline nickel, *Mater. Sci. Eng.*, **A301**, 35–43 (2001).
[37]Q.H. Bui, Polycristaux à grains ultrafins élaborés par métallurgie des poudres: microstructure, propriétés mécaniques et modélisation micromécanique," *PhD Thesis* University of Paris 13, Villetaneuse, (2008).
[38]Q.H. Bui, G. Dirras, S. Ramtani and J. Gubicza, On the strengthening behavior of ultrafine-grained nickel processed from nanopowders, *Mater. Sci. Eng.*, **A527**, 3227–3235 (2010).
[39]N. Krasilnikov, W. Lojkowski, Z. Pakiela and R. Valiev, Tensile strength and ductility of ultra-fine-grained nickel processed by severe plastic deformation, *Mater. Sci. Eng.*, **A397**, 330–337(2005).
[40]L. Farbaniec, Deformation mechanisms and fracture strength of polycrystalline ultrafine-grained materials: experimental and numerical investigations, *PhD Thesis*, University of Paris 13, Villetaneuse (2012).

[41]Brevet N°E.N. 07/58128 – 8 Octobre 2007, Utilisation de la technique de frittage flash pour la synthèse et la densification d'iodoapatites, L. Campayo (CEA ValRhô), S. Le Gallet (ICB), F. Bart (CEA ValRhô), F. Bernard (ICB), Y. Grin (Max Planck Institut, Dresde)
[42]L. Campayo, S. Le Gallet, Yu. Grin, F. Bart and F. Bernard, Sintering of lead phospho-vanadate by spark plasma sinterin, *J. Eur. Ceram. Soc.* **29**, 1477 (2009).
[43]L. Campayo, E. Courtois, S. Le Gallet, Yu. Grin and F. Bernard, Spark Plasma Sintering of Iodine-Bearing Apatite, J. *Nuclear Mater.* **400**, 251 (2010).
[44]M Gasik, Principles of Functionally Gradient Materials and their Processing by Powder Metallurgy, *Acta Polytechnica Scandinavica* **226** (1995)
[45]B Kieback, A.Neubrand and H.Riedel, Processing techniques for functionally graded materials, *Mater. Sci. Eng.,* **A362**, 81-105 (2003).
[46]C. Morin, Matériaux denses à gradient de fonction et à microstructure contrôlée préparés par SPS, *Ph-D Thesis*, Université de Bourgogne (2012).

FABRICATION OF TRANSPARENT $MgAl_2O_4$ SPINEL BY OPTIMIZING LOADING SCHEDULE DURING SPARK-PLASMA-SINTERING

Koji MORITA[1]*, Byung-Nam KIM[1], Hidehiro YOSHIDA[1], Yoshio SAKKA[1], Keijiro HIRAGA[2]
[1] *Advanced Ceramics Group, National Institute for Materials Science (NIMS), 1-2-1 Sengen, Tsukuba, Ibaraki 305-0047, Japan*
[2] *Materials Science and Engineering, Kitami Institute of Technology, 165 Koencho, Kitami, Hokkaido, 090-8507, Japan*
* Corresponding Author: *MROITA.Koji@nims.go.jp*

ABSTRACT

Loading schedule during spark-plasma-sintering (SPS) processing strongly affects the density and transmission of $MgAl_2O_4$ spinel. By loading at high temperatures, spinel can attain reasonable high density and resultant light transmission even at the high heating rate of 100°C/min though the transmission is lower than that attained by the low heating (<10°C/min) SPS processing. This suggests that during SPS processing, load controlling is an important factor as well as the heating rate and sintering temperature. In order to apply the high heating rate, which is the primary advantage of the SPS processing, the loading schedule should be optimized on the bases of the rate-controlling mechanisms of the densification. Carbon contamination, which is caused depending on the SPS process, also seems to be related to the limited transmission, which is lower than the highest value reported by HP technique.

INTRODUCTION

Spark-plasma-sintering (SPS) technique has widely been applied to the powder densification of various types of materials, such as metals, oxide and non-oxide ceramics, and composite materials [1-3]. The primary advantage of the SPS technique is available high heating rates of >50°C/min, which is much higher than that (typically ~10°C/min) of hot pressing (HP) or hot isostatic pressing (HIP) techniques. Due to the high heating rates, the SPS technique can save the processing time as compared with the well-known HP and HIP techniques. Hence, this method has been expected to be one of the realistic techniques for industrial applications.

The densification behavior, on the other hand, is known to be highly sensitive to the sintering parameters, such as heating rate, temperature, dwelling time and so on. For the synthesis of transparent alumina and spinel, for example, a low heating rate of $\alpha \leq 10$°C/min is highly effective than the widely used high heating rates of $\alpha \geq 50$°C/min during the SPS processing [4-11]; for $\alpha = 100$°C/min, spinel is opaque, whereas for $\alpha \leq 10$°C/min, it shows in-line transmission of ≈47% at the visible range. Although the low heating rate SPS processing can attain good light transmission in spinel, it needs a long processing time as well as the HP or HIP techniques. This is a negative point of the low heating rate SPS processing. In order to fully utilize the primary advantage of the SPS technique, the optimum high heating rate processing should be established to save the processing time.

According to the previous studies [12-15], controlling the loading schedule seems to be one of the remaining factors and a candidate to develop the high heating rate SPS processing. The effect of the loading schedule has already been examined in the several transparent oxide ceramics, such as MgO [12], Al_2O_3 [13], spinel [14] and Lu_2O_3 [15]. In those ceramics, the transmission can be improved using the two-step loading method, in which the applied load was maintained at smaller levels during the heating process and increased to higher levels at arbitrary temperatures [12-15]. In order to develop the optimum high heating rate SPS processing for the fabrication of highly dense ceramics, although understanding the loading effects on the densification behavior would be important, the detailed densification mechanism seems to remain unsolved.

Another problem is that the transmission of SPSed spinel is lower than that of HPed one. This seems to be related to discoloration, which is a typical phenomenon in the SPS processed materials [10]. Although the reason is not clear at the present time, the discoloration is remarkable than HP processed materials. In order to attain much higher transmission, understanding the source of discoloration would be important

Therefore, this study was carried out to examine the loading effect on the transmission and on the microstructures of spinel polycrystals.

EXPERIMENTAL PROCEDURES

A high purity $MgAl_2O_4$ spinel powder (TSP-15, Taimei Chemical Co., Ltd., Tokyo, Japan) was sintered with a spark-plasma-sintering machine (SPS-1050, SPS Syntex, Inc., Kawasaki, Japan) under vacuum condition. The details were described elsewhere [16].

In briefly, the as-received spinel powder has a purity of >99.97% and a particle size of 100-300nm. The powder was placed in a graphite die with a 30 mm inner diameter. Under vacuum condition of about 10^{-3} torr, the sintering was carried out at 1300°C for a 20min soak and at various heating rates ranging from α = 10°C/min to 100°C/min. In order to examine the effect of the loading schedule on the sinterability of the spinel powder, a sintering load of 80MPa was applied under several conditions. Prior to the heating, an available small load of about 5MPa was applied to the graphite die to maintain current conductivity. During the heating process, a sintering load of 80MPa was rapidly applied within 10s when the temperature reaches arbitrary temperatures. During the SPS processing, the temperature was measured on the surface of the graphite die using an optical pyrometer.

For the optical characterization, square plates with the dimensions of 12×12mm were machined from the center of the sintered circular disks with a 30 mm diameter. Both surfaces of the plates were mirror-polished using 1 μm diamond paste and the thickness of the square plate was reduced to about 1.8 mm. The in-line transmission T_{in} was measured using a double-beam spectrophotometer (SolidSpec-3700DUV, Shimazu Co., Ltd., Kyoto, Japan) in the visible- and near-IR-wavelength range (λ = 240-2200 nm).

The microstructures were also characterized near the center of the sintered circular disks with a scanning electron microscope (SEM : JSM-6500F, JEOL Co., Ltd., Tokyo, Japan). The surface of the specimen was mechanically mirror-polished with 1 μm diamond paste. In order to reveal the grain boundaries, the polished specimen was slightly thermal-etched at 1200°C for 30 min in air. The grain size was determined by counting the number of grains. Assuming the grains to be spherical shape, the average grain size, $\bar{d}$, was determined to be 1.225 times the apparent grain size, which was calculated from the average cross sectional area per grain [17]. The residual pore density, N, which was defined as the number of pores per unit area, was measured by counting the number of pores using the SEM images.

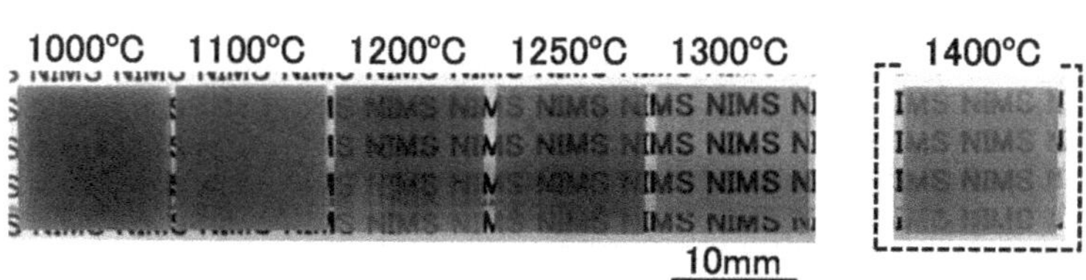

Figure 1. Spinel plates SPS processed at 1300°C for a 20min soak and at α =100°C/min [16]. A sintering load of 80MPa was applied at the arbitrary temperatures ranging from 1000°C to 1300°C. For comparison, the sintering load of 80MPa was applied at 1400°C for 20min.

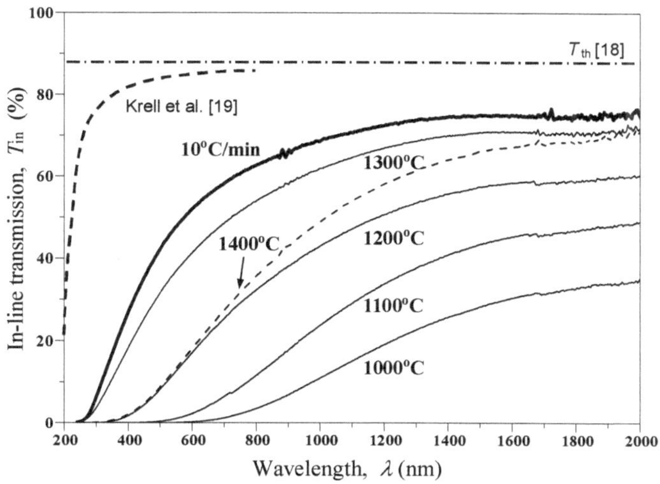

Figure 2. Load temperature dependent in-line transmission T_{in} vs. wavelength λ [16]. The theoretical transmission, T_{th}, expected for a single $MgAl_2O_4$ spinel crystal [18] is shown by dash-dotted line. The highest T_{in} [19-21], which is HIP treated at 1360°C, are also shown by broken line for comparison. The data were normalized at the same thickness of 1.8mm.

In order to examine the residual carbon contamination after the SPS processing, the sintered spinel was characterized with a Raman spectroscopy (T64000, Horiba Co., Ltd., Tokyo, Japan) equipped with an integrated triple spectrometer system. After removing the surface of the sintered spinel, Raman spectrum was carefully measured using green leaser of λ = 514.532nm for 400s per each measurement.

RESULTS AND DISCUSSION

Figure 1 shows the photo of the mirror-polished spinel plates [15], in which a sintering load of 80MPa was applied at several temperatures to examine the effect of the loading temperature on the transmission. It is clear from the photo that although the sintering temperature and duration time are the same (1300°C and 20min) in all the samples, its transmission changes sensitively with the loading temperature. For the loading at $T \leq$ 1100°C, the spinel plates are almost opaque, whereas at $T \geq$ 1200°C, they show transmission. The text placed 10 mm below the plates becomes clear as the loading temperature increases. In the case of the higher temperature loading at 1400°C for a 20min soak, however, the transparency seems to slightly decrease as compared with that of 1300°C loading. It should also be noted that the discoloration of the spinel plate tends to be pronounced with loading temperature.

Figure 2 represents the in-line transmission, T_{in}, of each spinel plate shown in Fig. 1 [15]. The transmission of spinel depends strongly on the loading temperature during SPS processing. For the 1300°C loading, spinel shows good transmission, irrespective of the high heating rate of α = 100°C/min. As compared with the highest T_{in}-value of >80% reported in the HPed spinel by Krell *et al.* [19-21], the T_{in}-value is much lower. However, T_{in} can clearly be improved in the wide wavelength range with increasing the loading temperature and the T_{in}-value of the 1300°C loading becomes comparable to the highest value (>80% [19]).

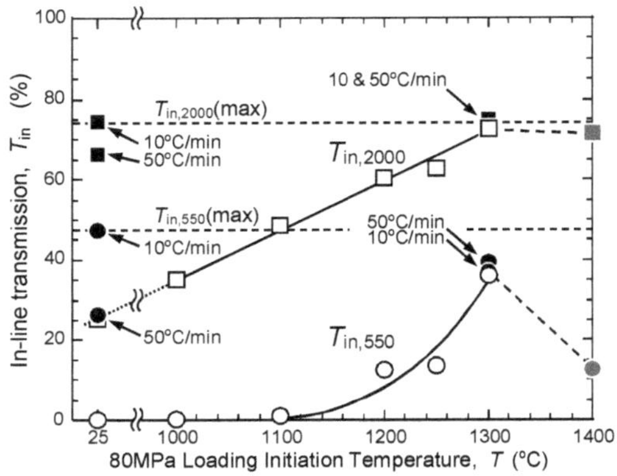

Figure 3. $T_{in,550}$ and $T_{in,2000}$ plotted as a function of load temperature [16]. The maximum in-line transmission ($T_{in,550}$(max) and $T_{in,2000}$(max)), which are attained by the low heating rate SPS processing at 10°C/min and at a 20min soak at 1300°C [8-11], are shown by dash lines. The closed symbols are T_{in} obtained at different heating rates of 10 and 50°C/min.

In order to make clear the influence of the loading temperature on the transmission, the in-line transmission evaluated at the visible- and infrared-wavelengths of λ=550 and 2000 nm ($T_{in,550}$ and $T_{in,2000}$) are plotted in Fig. 3 as a function of the loading temperature [15]. For comparison, the maximum $T_{in,550}$- and $T_{in,2000}$-values attained by the low heating rate SPS processing at α = 10°C/min and at 1300°C [8-10] are also indicated by $T_{in,550(max)}$ and $T_{in,2000(max)}$, respectively. The $T_{in,2000}$-value can be improved with the loading temperature and reaches ≈74% at the 1300°C loading, which is almost the same to $T_{in,2000(max)}$. On the other hand, the $T_{in,550}$-value also increases with the loading temperature at above 1100°C and exhibits the highest value of ≈35% at the 1300°C loading. Although this visible light transmission of ≈35% is lower than $T_{in,550(max)}$ (≈47%), it can effectively be improved as compared with that ($T_{in,550}$ ≈ 0%) of the 80MPa pre-loading prior to the heating. This suggests that the high heating SPS processing can also be applied to the fabrication of transparent ceramics if the sintering load was controlled, and the high temperature loading would be effective for attaining high transmission.

However, much higher temperatures are not necessarily effective for attaining the transparency. For the loading at the higher temperature of 1400°C, although the transmission is less sensitive at $T_{in,2000}$, it deteriorates at $T_{in,550}$ as shown in Figs. 2 and 3, suggesting that an optimum sintering temperature seems to exist. For simultaneously attaining high transmission in the wide wavelength range from visible to infrared lights, the loading at 1300°C is very efficient for the sintering of the present spinel powder.

This loading temperature dependent transmission shows a trend similar to the previous studies [12-15], in which the transmission of oxide ceramics can also be improved by the two-step loading method. For the sintering of spinel, for example, Wang and Zhao [14] reported that high transparent spinel can be attained by the two-step loading method at 1300°C sintering for 3 min soak even at the high heating rate SPS process of $\alpha \geq$ 50°C/min. They applied several pre-loads ranging from 5 to 100MPa during the heating process and changed the load to 100MPa when the temperature reached 1250°C. By employing this two-step loading method, they attained a transparent spinel of $T_{in,550}$ ≈ 50% when they employed a smaller pre-load of 5MPa.

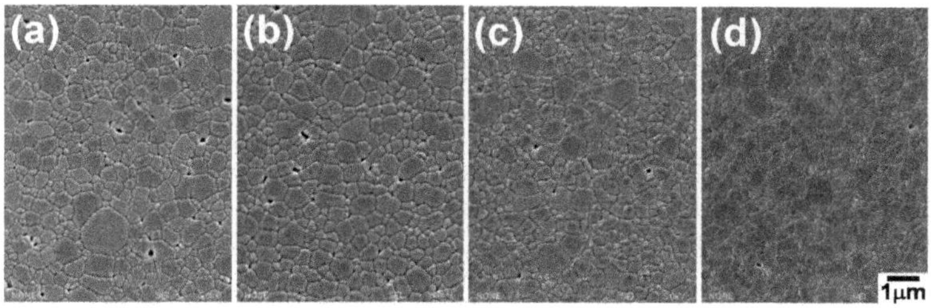

Figure 4. Typical SEM images of spinel SPSed at a 20min soak at 1300°C with a high heating rate of α = 100°C/min [16]. During the heating-up process, the sintering load of 80MPa was rapidly applied at (a) 1000, (b) 1100, (c) 1200 and (d) 1300°C.

In order to confirm the effect of the loading temperature on the transmission, the microstructures of the sintered spinel were examined because transmittance is highly sensitive to the microstructural factors, such as residual porosity and impurities.

Figure 4 shows typical microstructures of spinel sintered at several loading schedules [15]. Wang and Zhao [14] reported that the two-step loading method in the SPS process does not affect the microstructure of spinel. The grain size is indeed independent of the loading temperature as well as their result. However, in contrast to their explanation, the residual porosity is strongly influenced by the loading temperature as shown in Fig. 4. The loading temperature dependent porosity N consists well with the experimental results reported in other oxide ceramics [13-15]. For the loading at lower temperatures of ≤1100°C, many closed pores larger than 100 nm are observed to remain at the multiple grain junctions (Figs. 4(a) and (b)), and the pore density tends to decrease with increasing the loading temperature (Figs. 4(c) and (d)). The pore density N is plotted as a function of the loading temperature in Fig. 5 [16]. For the loading at the high temperature of 1300°C, although nanometer-sized fine pores are still observed to remain in the limited multiple grain junctions as shown in Fig. 4(d), the pore density can be reduced to the lower level, which is similar to the residual pore density attained by the low heating rate SPS processing [8-11].

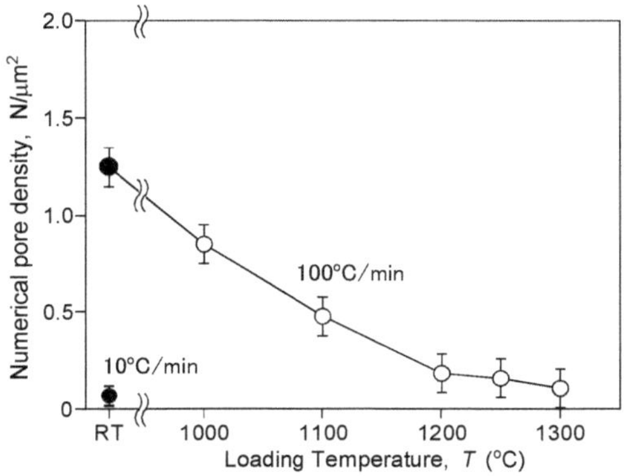

Figure 5. Pore density N plotted as a function of the loading temperature [16].

The light scattering by the residual pores is known to have a serious influence on the light transmission from the visible through infrared wavelengths [4-11,17]. The loading temperature dependence of both the light transmission and the residual porosity exhibits a similar trend to each other. Wang and Zhao [14] suggested that the pre-loading dependent transmission can be explained by the absorption caused by the oxygen vacancies, which were generated depending on the pre-loading level [14]. However, the absorption caused by the oxygen vacancies is known to be remarkable in the UV range of λ < 400nm [22,23], and hence, cannot explain the transmission improved in the wide wavelength range from visible to IR light (Fig. 2). For the present study, the oxygen vacancy is unlikely to be the dominant reason for the present loading temperature dependence though it may affect the limited transmission at the shorter wavelength range. It is reasonable to conclude based on the experimental facts that the effect of the loading temperature on the transmission can be mainly ascribed to the light scattering caused by the residual pores.

Another interesting point is that if the load was applied at the high temperature of 1300°C, the light transmission of spinel is insensitive to the heating rate, particularly in the visible ranges, as shown in Fig. 3. We reported that when the larger load of 80MPa was applied at room temperature (≈25°C) prior to the heating process, transmission sensitively changes with the heating rate during the SPS processing and $T_{in,550}$ can be improved from 0% at α = 100°C/min to ≈47% at α = 10°C/min, as indicated by the closed symbols in Fig. 3. For the rapid loading at the high temperature of 1300°C, however, both the $T_{in,550}$- and $T_{in,2000}$-values exhibit almost the same values for all the heating rates (Fig. 3), though the $T_{in,550}$-value is lower than $T_{in,550(max)}$. This indicates that the loading condition strongly affects the densification of the spinel rather than the heating rate during SPS processing.

This suggests that load-related densification mechanisms during the SPS processing are likely to be a main factor of the loading temperature dependent densification. If the larger sintering load applies prior to the heating process, the strong inter-particle bonds enhance the formation of large closed pores between the particles, as noted in a previous study [9]. In order to attain dense bulks, pores should be shrunk by diffusions through the continuous pore channel during the initial and intermediate sintering stages [24]. Once the large closed pores are formed due to the inter-particle bonds, they are difficult to remove in the later sintering stages and those would remain at the multiple grain junctions. On the other hand, for the 80MPa loading at high temperatures, the densification must smoothly take place through the plastic deformation and diffusion processes [25], and thereby, the pores may shrink through the continuous pore channel without forming the closed pores. Therefore, the high densities can be attained independent of the heating rates. Consequently, in order to optimize the SPS processing, the loading schedule should be controlled on the bases of the rate-controlling mechanisms of the densification.

Although the transmission of spinel can be improved by the high temperature loading, it is still much lower than the highest value attained by the HP technique [19-21]. Since the residual porosity can be reduced by controlling the loading schedule, other factors may affect the light transmission of the SPSed spinel. Figure 6 shows Raman spectra obtained from the spinel sintered at low and high temperature loading conditions [26]. The Raman spectra clearly reveal two broad peaks at around 1350 cm^{-1} and 1600 cm^{-1}, which are related to the D- and G-bands arising from glassy carbon. The carbon peak is slightly sharp and larger in low temperature loading than in high temperature loading. This suggests that the carbon contamination during SPS processing seems to be enhanced in the low temperature loading condition. Although careful examinations are necessary to discuss the effect of the trace carbon contamination on the light transmission of SPSed materials, the lower transmission and discoloration in the SPSed materials, particular in the low temperature loading, may be ascribed to the scattering and/or absorption caused by the carbon contamination.

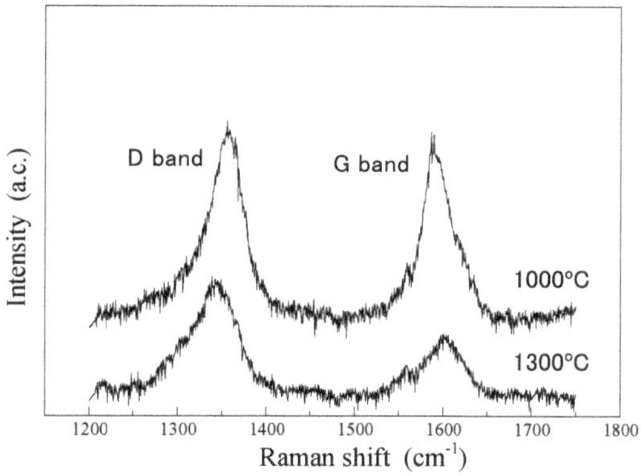

Figure 6. Raman spectra of spinel loaded at 1000°C (upper) and 1300°C (bottom) [26].

CONCLUSION

The effect of the loading temperature during the SPS processing was examined in $MgAl_2O_4$ spinel. An increasing the loading temperature can improve the light transmission of spinel. Although the transmission is lower than those ($T_{in,550} \approx 47\%$ and $T_{in,2000} \approx 74\%$) attained by the low heating rate SPS processing, the loading at a high temperature of 1300°C can attain a good light transmission of $T_{in,550} \approx 35\%$ and $T_{in,2000} \approx 74\%$ in polycrystalline spinel even for the high heating rate SPS processing of 100°C/min. Since the high temperature loading can reduce the residual pores, which is a main light scattering source, the loading temperature dependent T_{in} can be mainly explained by the scattering by the residual pores. For the high temperature loading, the densification smoothly takes place through the plastic deformation and diffusion, and hence, the high densities can be attained. The larger pre-loading at lower temperatures, on the other hand, enhances the formation of strong inter-particle bonds and results in the enhanced formation of large closed pores. The loading temperature also affects on the carbon contamination. It can be concluded from the Raman measurements that the low temperature loading tends to increase the carbon contamination and this deteriorates the light transmission of SPSed spinel.

ACKNOWLEDGEMENTS

Authors would like to appreciate sincerely to the financial supports by a Grant-in-Aid for Scientific Research (C) 22560675 and (B) 21360328 from the Ministry of Education, Culture, Sports, Science and Technology (MEXT), Japan.

REFERENCES

1 Z. A. Munir, U. Anselmi-Tamburini, and M. Ohyanagi, The effect of electric field and pressure on the synthesis and consolidation of materials: A review of the spark plasma sintering method, *J. Mater. Sci.*, **41**, 763-77 (2006).
2 R. Orrù, R. Licheri, A. M. Locci, A. Cincotti, and G. Cao, Consolidation/synthesis of materials by electric current activated/assisted sintering, *Mat. Sci. Eng. R*, **63**, 127-87 (2009).
3 S. Grasso, Y. Sakka, and G. Marizza, Electric current activated/assisted sintering (*ECAS*): a review of patents 1906–2008, *Sci. Technol. Adv. Mater.*, **10**, 053001-24 (2009).
4 B.-N. Kim, K. Hiraga, K. Morita, and H. Yoshida, Spark plasma sintering of transparent alumina, *Scripta Mater.*, **57**, 607-10 (2007).

5 B.-N. Kim, K. Hiraga, K. Morita, and H. Yoshida, Effects of heating rate on microstructure and transparency of spark-plasma-sintered alumina, *J. Eur. Ceram. Soc.*, **29**, 323-27 (2009).
6 B.-N. Kim, K. Hiraga, K. Morita, and H. Yoshida, T. Miyazaki, and Y. Kagawa, Microstructure and optical properties of transparent alumina, *Acta Mater.*, **57**, 1319-26 (2009).
7 B.-N. Kim, K. Morita, J.-H. Lim, K. Hiraga, and H. Yoshida, Effects of preheating of powder before spark plasma sintering of transparent $MgAl_2O_4$ spinel, *J. Am. Ceram. Soc.*, **93**, 2158-60 (2010).
8 K. Morita, B.-N. Kim, K. Hiraga, and H. Yoshida, Fabrication of transparent $MgAl_2O_4$ spinel polycrystal by spark-plasma-sintering processing, *Scripta Mater.*, **58**, 1114-17 (2008).
9 K. Morita, B.-N. Kim, K. Hiraga, and H. Yoshida, Spark-plasma-sintering condition optimization for producing transparent $MgAl_2O_4$ spinel polycrystal, *J. Am. Ceram. Soc.*, **92**, 1208-16 (2009).
10 K. Morita, B.-N. Kim, K. Hiraga, and H. Yoshida, Fabrication of high-strength transparent $MgAl_2O_4$ spinel polycrystals by optimizing spark-plasma-sintering conditions, *J. Mater. Res.*, **24**, 2863-72 (2009).
11 K. Morita, B.-N. Kim, H. Yoshida, Z. Haibin, K. Hiraga, and Y. Sakka, to be submitted.
12 R. Chaim, Z. Shen, and M. Nygren, Transparent nanocrystalline MgO by rapid and low-temperature spark plasma sintering, *J. Mater. Res.*, **19**, 2527-31 (2004).
13 S. Grasso, C. Hu, G. Maizza, B.-N. Kim, and Y. Sakka, Effects of pressure application method on transparency of spark plasma sintered alumina, *J. Am. Ceram. Soc.*, **94**, 1405-09 (2010).
14 C. Wang, and Z. Zhao, Transparent $MgAl_2O_4$ ceramics produced by spark plasma sintering, *Scripta Mater.*, **61**, 193-96 (2009).
15 L. An, A. Ito, T. Goto, Two-step pressure sintering of transparent lutetium oxide by sparking plasma sintering, *J. Eur. Ceram. Soc., J. Eur. Ceram. Soc.*, **31**, 1597-1602 (2011).
16 K. Morita, B.-N. Kim, H. Yoshida, H. Zhang, K. Hiraga, and Y. Sakka, Effect of Loading Schedule on Densification of $MgAl_2O_4$ Spinel during Spark Plasma Sintering (SPS) Processing, J Eur. Ceram. Soc., 32, 2303-09 (2012).
17 R. Apetz, and M. P. B. van Bruggen, Transparent alumina: a light-scattering model, *J. Am. Ceram. Soc.*, **86**, 480-86 (2003).
18 A. F. Dericioglu, and Y. Kagawa, Y., Effect of grain boundary microcracking on the light transmittance of sintered transparent $MgAl_2O_4$, *J. Euro. Ceram. Soc.*, **23**, 951-59 (2003).
19 A. Krell, J. Klimke, and T. Hutzler, Transparent compact ceramics: inherent physical issues, *Optical Mater.*, **31**, 1144-50 (2009).
20 A. Krell, J. Klimke, and T. Hutzler, T., Advanced spinel and sub-μm Al_2O_3 for transparent armour applications, *J. Eur. Ceram. Soc.*, **29**, 275-81 (2009).
21 A. Krell, T. Hutzler, J. Klimke, and A. Potthoff, Fine-grained transparent spinel windows by the processing of different nanopowders, *J. Am. Ceram. Soc.*, **93**, 2656-66 (2010).
22 L. S. Cain, G. J. Pogatshnik, and Y. Chen, Optical transitions in neutron-irradiated $MgAl_2O_4$ spinel crystals, Phys. Rev. B, **37**, 2645–52 (1988).
23 J. He, L.-B. Lin, T.-C. Lu, and P. Wang, Effects of electron- and/or gamma-irradiation upon the optical behavior of transparent $MgAl_2O_4$ ceramics: Different color centers induced by electron-beam and γ-ray, Nucl. Instr. Meth. Phy. Res. B, **191**, 596-99 (2002).
24 M. N. Rahaman, Sintering of Ceramics, CRC Press, Boca Raton, FL (2008).
25 K. Morita, B.-N. Kim, H. Yoshida, and K. Hiraga, Densification behavior of a fine-grained $MgAl_2O_4$ spinel during spark-plasma-sintering (SPS), *Scripta Mater.*, **63**, 565-68 (2010).
26 K. Morita, B.-N. Kim, H. Yoshida, and K. Hiraga, and Y. Sakka, to be submitted.

PROPERTIES OF WCCO/DIAMOND COMPOSITES PRODUCED BY PPS METHOD INTENDED FOR DRILL BITS FOR MACHINING OF BUILDING STONES

Marcin Rosinski*, Joanna Wachowicz, Tomasz Plocinski, Tomasz Truszkowski, Andrzej Michalski

Warsaw University of Technology, Faculty of Materials Science and Engineering
141 Woloska Str.,02-507 Warsaw, Poland
*ninmar@inmat.pw.edu.pl

ABSTRACT

The paper presents the application of the pulse plasma sintering (PPS) method in the field of diamond composites sintered under the conditions of thermodynamic instability of diamond for the manufacture of tools intended for machining building stone. The WCCo/diamond composites containing 30 vol % of diamond particles were produced using a mixture of submicron WC6Co (wt %). Thanks to PPS densification conditions, dense sinters with a strong bond between the diamond particles and the sintered carbide matrix have been obtained. Examinations of the phase composition and observations of the microstructure did not show graphitization of diamond. The SEM photographs revealed transcrystalline fractures of the diamond particles. The presence of transcrystalline fractures of the diamond particles indicates that the bonding forces between the diamond particles and the WCCo matrix exceed the strength of the diamond particles. The paper compares the percent number of transcrystalline fractures of diamond particles in dependence on the densification parameters.

INTRODUCTION

Diamond is a metastable phase and under normal pressure and at high temperatures it is transformed into graphite. To avoid this transformation, polycrystalline diamond (PCD) is sintered within the temperature range from 1500 to 2000°C under a pressure of 4-5 GPa. Because of the necessity of employing the expensive HPHT (High Pressure-High Temperature) sintering technique, the price of these tools is high. In turn, when the sintering process is to be conducted at normal pressure, the refractory materials such as sintered carbides are rarely used as the matrix of diamond cutting tools since they must be maintained at a high temperature for a long time. Usually, sintered carbide is sintered with the participation of melted cobalt. However, as mentioned earlier, at normal pressure and at the sintering temperature of WCCo (1400-1500°C), diamond is a metastable phase and undergoes graphitization. Therefore, the WCCo composites, containing diamond particles distributed within it, must not be sintered under these conditions. The present study was concerned with the production of diamond sinters with a WCCo-based matrix under the condition of thermodynamic instability of diamond.

In the last decade SPS (Spark Plasma Sintering) have been considered to belong to the most efficient and most significant densification methods [1-5]. Just as in conventional hot-pressing (HP) the densification process is here conducted under pressure, but the essential difference lies in the way in which the thermal energy is transferred to the material being sintered. In conventional sintering, the thermal energy is delivered through radiation and heat conduction so that heat is transferred from the surface of the material to its core and, thus, the heating rate is low and the heating efficiency is poor. In the SPS methods, the thermal energy is dissipated directly within the entire volume of the material, thanks to which the energy losses to the environment are small and the energy consumption is low.

H. Moriguchi et al. [6] sintered diamond/cemented carbide composites by the SPS method at a temperature of 1300°C under a pressure of 41 MPa for 3 min. To avoid the graphitization of diamond during the densification, they covered the diamond particles with a SiC layer. X. L. Shi et al. [7], who sintered diamond/cemented carbide by SPS at a temperature between 1000 and 1280°C, covered the diamond particles with a tungsten layer.

The paper is concerned with the application of the pulse plasma sintering (PPS) method to densification, under the conditions of thermodynamic instability of diamond, new diamond composites intended for the manufacture of tools suitable for machining building stone. The PPS method is an original technique of pressing powders. The innovatory idea is here the use of electric current impulses, with the amplitude of the order of several hundred kA, which are generated by discharging a capacitor battery. Thus far, the PPS method has been used for densification a wide variety of materials, such as nanocrystalline materials [8-10], WC/Ti/Co-cBN [11-12], WCCo/diamond [13], and Cu/diamond [14] composites.

EXPERIMENTAL METHODS

The WCCo/diamond composite containing 30 vol % of diamond particles was produced using a mixture of submicron WC6Co (wt %) added with an MBD4 diamond powder (particle size – 16 - 20 μm) delivered by the Luoyang High-Tech Qiming Superhard Materials Co. The mixture was prepared in a turbular mixer and the mixing operation lasted for 10 h. The samples were sized at 20 mm in diameter and 6 mm in height and were sintered in a graphite die at a heating rate of 500°C/min. The densification process was conducted using the PPS method under a load between 60 and 100 MPa at a pressure of 5.10^{-3} mbar and a temperature of 1050-1100°C for 5 min.

The microstructure of the sintered samples was observed in scanning electron microscope (SEM), and their phase composition was examined using a Philips PW 1140 diffractometer. The density of the sintered samples was measured by the Archimedes method. The hardness was measured at room temperature by the Vickers diamond indentation method using a ZWICK hardness-meter under a load of 1 kG. The WC particle size was determined based on the SEM photographs of the microstructures of the composite fractures by the MicroMeter program (developed at the Warsaw University of Technology, Faculty of Materials Science and Engineering [14,15]) which calculated the sizes of the individual grains. The percentage share of the transcrystalline fractures of the diamond particles present in the WCCo/diamond composite was determined with the use of the MicroMeter program.

Because of the differences in the sputtering rates between WC, Co and diamond it is very difficult to prepare electron-transparent TEM specimens using conventional methods. In order to characterize precisely the structure of the WCCo/diamond transition layer in a desired area, and taking into account the differences between the properties of WC, Co and diamond, the samples had to be specially prepared using the focused ion beam (FIB) technique so as to make them suitable for examination in a scanning transmission electron microscope (STEM). Figure 1 shows the surface of a WCCo/diamond composite after grinding and polishing. In order to achieve a relatively flat surface suitable for thinning by the FIB technique, the material was polished with diamond-embedded resins. Microstructure investigations were performed on the cross-section sample cutted across the interface between WCCo matrix and diamond particle. A single beam Hitachi FB 2100 FIB was used for imaging the surface of the sample and localization of that interface. The sample was prepared by using a Ga^+ ion beam at a voltage of 40 kV. The microstructure investigations of the prepared sample were performed on a SEM HITACHI S-5500 equipped with DUO-STEM detector and EDS-Thermo Noran VANTAGE system for chemical analyses.

RESULTS

The density of the WCCo/diamond composite sintered at a temperature of 1050°C under a load of 100 MPa for 5 min. was 99.9 % of the theoretical density (TD - 11.38 g/cm^3 - determined from the mixing rule) whereas the density of the composite sintered at the same temperature but under a load of 60 MPa was only 11.07 g/cm^3 (97.3 % TD). The density of the WCCo/diamond composite sintered at a temperature of 1100°C under a load of 60 MPa for 5 min increased to 98.7 % of TD. Hence we can see that any increase of the load results in an increase of the density of the composite irrespective of the densification temperature.

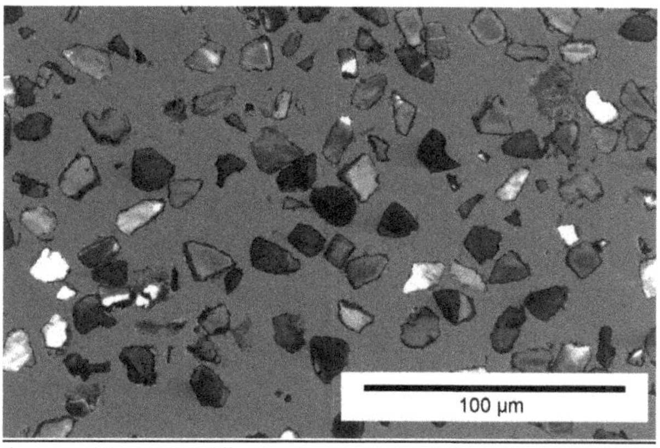

Fig. 1. SEM image of the WCCo/diamond composite surface after grinding and polishing.

Figure 2 shows a diffractogram of the WCCo/diamond composite sintered at a temperature of 1100°C. We can see that the composite contains the three phases: diamond, tungsten carbide and cobalt.

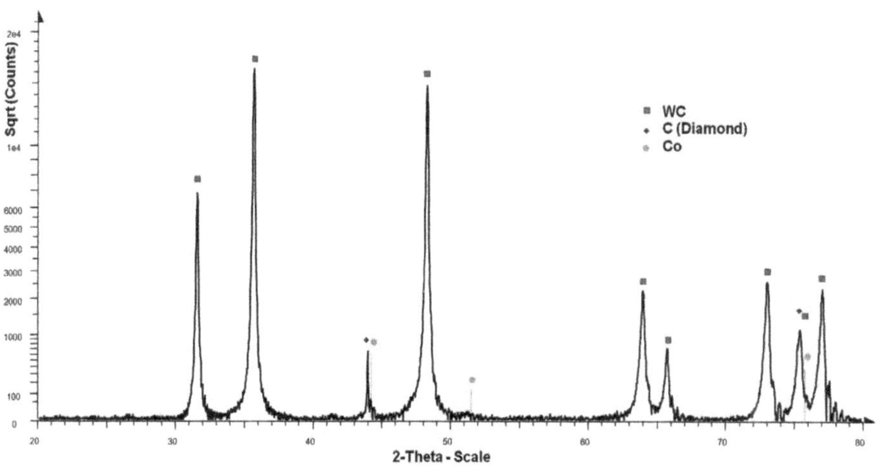

Fig. 2. Diffraction spectrum obtained for the WCCo/diamond composite sintered at 1100°C.

The hardness of the WCCo/diamond composite, containing 30 vol. % of diamond particles, sintered at a temperature of 1050°C under a load of 100 MPa for 5 min (Fig. 3) using a mixture of a WC powder (with a particle size of 0.4 μm) added with 6 wt % of Co and a diamond powder (particle size ranging from 16 to 20 μm) was higher by about 2 GPa than that of the cemented carbide with the same WC grain size and the same Co content.

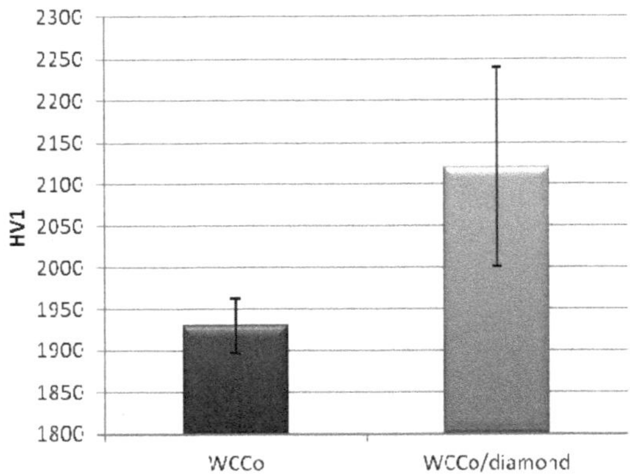

Fig. 3. The hardness of the cemented carbide with 6 wt % of Co and of the WCCo/diamond composite sintered at a temperature of 1050°C under a load of 100 MPa for 5 min.

Figure 4 shows a SEM image of the surface of a fracture in the WCCo/diamond composite matrix (composite sintered at a temperature of 1100°C under a load of 100 MPa). We can see that the diamond particles are well bound with the WCCo matrix, and no pores are present around them. On the composite fracture surface, the SEM photograph reveals small diamond particles (d), voids left by the diamond particles thorn out from the WCCo matrix (v), and transcrystalline fractures of the diamond particles (f).

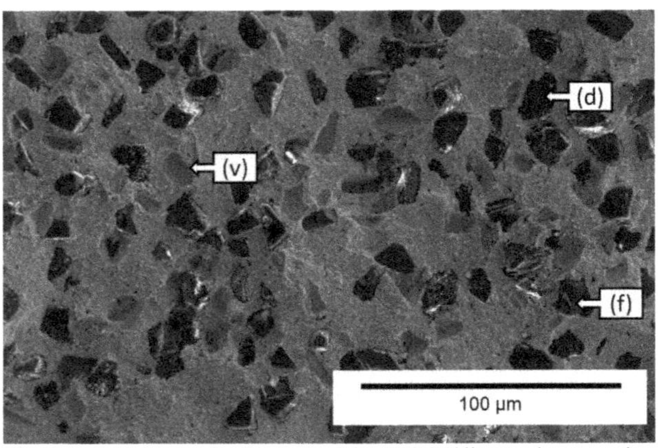

Fig. 4. SEM image of a fracture of the WCCo/diamond composite.

Figure 5 compares the percent number of these transcrystalline fractures (black), the number of diamond particles (dotted), and the number of the voids left by the thorn-off diamond particles (gray) in dependence on the densification temperature (within the 1050-1100°C range). The numbers of the unbroken diamond particles (denoted as (d) in Fig. 4) and of the voids left by the thorn-off diamond particles (denoted as (v) in Fig. 4), regardless of the densification temperatures (1100, 1075 and 1050°C) were statistically about the same (48(d) – 43(v), 47(d) – 44(v) and 43(d) – 40(v) %, respectively). The decrease of the densification temperature from 1100 to 1050°C increased the number of the transcrystalline fractures of the diamond particles more than twice (from 8 to 17 %). The presence of transcrystalline fractures of the diamond particles indicates that the bonding forces between the diamond particles and the WCCo matrix exceed the strength of the diamond particles.

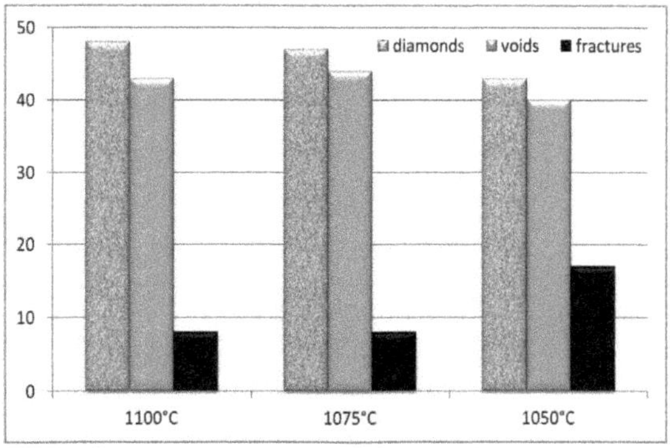

Fig.5. Effect of the densification temperature on the number of transcrystalline fractures (in black), diamond particles (dotted), and the voids left by the diamond particles thorn out from the composite matrix (gray).

This strong bond can be achieved thanks to the formation of a transition layer at the matrix/diamond interface. Figure 6 is a BF STEM photograph of the WCCo/diamond composite sintered at a temperature of 1050°C under a load of 100 MPa for 5 min. We can see a fragment of a diamond particle and of a WC grain with the transition layer formed in-between them. The STEM analyses of the WCCo/diamond interface revealed that: the diamond particles were well bonded to the WCCo matrix, no interfacial voids or debonding occurred and an interfacial layer had formed between a diamond particle and the WCCo matrix. The WCCo matrix well adhering to diamond particles what can be observed on the Fig. 6. The STEM analyses also revealed a well-defined discontinuous layer about 50-300 nm thick located at the interface between the WCCo matrix and the diamond particle surface.

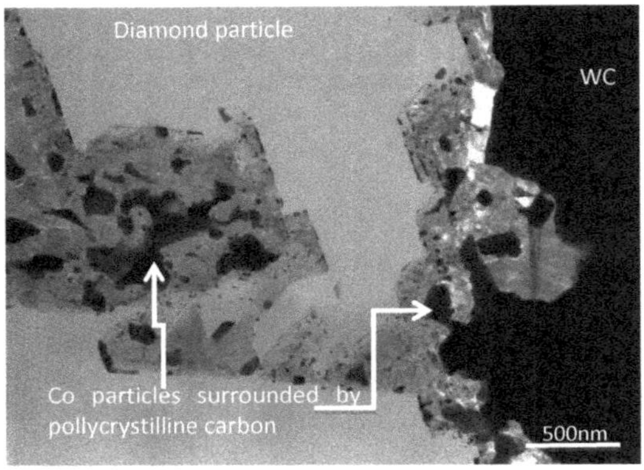

Fig. 6. BF STEM photograph of the WCCo/diamond composite.

The bond between the diamond and the WCCo matrix in the WCCo/diamond composite appeared to be strong which was due probably to the formation of a transition layer composed of a solid solution of tungsten and carbon in cobalt. An EDS analysis of these regions (Fig. 7) confirmed the presence of cobalt and tungsten there.

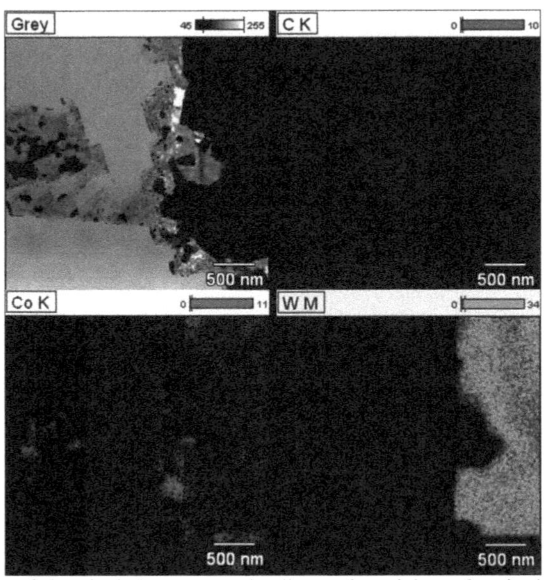

Fig. 7. Distribution of chemical elements on the diamond particle surface in the WCCo/diamond composite.

The EDS line scans through the interface confirmed these findings. The C, W, Co signals (Fig. 8), recorded along the line perpendicular to the WCCo/diamond boundary show an increase of the cobalt signal at the transition layer.

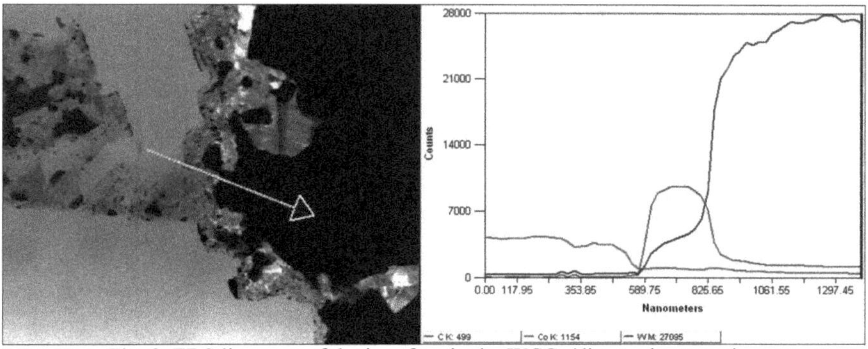

Fig. 8. EDS line scan of the interface in the WCCo/diamond composite.

The average WC grain size in the composite sintered at a temperature of 1050°C was 0.41 μm, i.e. it is comparable with the WC grain size (0.4 μm) in the starting WC powder. This means that, in this composite, the grains did not grow up during the densification process. After densification at a temperature of 1100°C, the WC grain size increased to 0.49 μm, but the grain size distribution remained unchanged. In both composites, most of the grain sizes fall within the range from 0.2 μm to 0.6 μm.

Figure 9 is a SEM photograph of the surface of a fracture of the matrix in the WCCo/diamond composite sintered at a temperature of 1050°C under a load of 100 MPa. We can see WC grains with well shaped crystallographic walls and cobalt paths formed in-between them. This indicates that, during the PPS process, the cobalt phase occurs in the melted state even when the WCCo/diamond composite is sintered at a temperature of 1050°C, i.e. below the eutectic temperature of this sinter.

Fig. 9. Fracture through the matrix of a WCCo/diamond composite (SEM image).

DISCUSSION

The occurrence of the liquid cobalt phase during pulse plasma sintering is associated with the rapid increase of the temperature up to several thousand Celsius' grades during the flow of the current pulse through the material, as was demonstrated in our earlier study devoted to the densification of the composites with a cemented carbide matrix and diamond particles dispersed within it [13]. This observation was also confirmed by the present experiments, since the WC grains visible in the microstructure of the cemented carbides PPS-sintered at a temperature of 1050°C (Figure 9) are evenly surrounded by the binding cobalt phase, which is evidence that the high-current pulse densification proceeds with the participation of this phase in the liquid state. When the sintering process runs without the participation of the liquid phase, cobalt usually occurs in the form of agglomerates and is distributed non-uniformly in the cemented carbide matrix [17]. In our experiments however the densification process was conducted at a temperature of 1050°C which is below the temperature at which the liquid cobalt phase should occur. According to ref. [18], in the WC-Co alloys the cobalt liquid phase appears at a temperature of 1220°C if the alloy contains 5 wt % Co, and at 1190°C in the alloy with 12 wt % Co. In the PPS method, the material to be sintered is heated by periodically repeated electric current pulses with the amplitude of about 60 kA and the pulse duration of about 0.5 ms. Just as in the FAST and SPS techniques, the surfaces of the material grains are heated to a very high instantaneous temperature of the order of several thousand Celsius' grades [7]. The essence of the PPS method is that the heating operation is realized by the application of high current pulses with the intensity of several tens of kA, obtained by discharging a capacitor battery. The use of capacitors as the source of the energy necessary for the consolidation of the powder creates specific heating and cooling conditions since the energy of several kJ is delivered to the processed powder during a time as short as several hundred microseconds. During the current flow, the powder being consolidated is heated to a high temperature and, after the current decays, the powder quickly cools down to the specified densification temperature.

Moreover, since the diamond particles do not conduct electric current, the current density in their vicinity increases, resulting in an increase of the amount of the dissipated Joule heat and, thus, the increase of the instantaneous temperature to above the melting point of cobalt. In effect, during the high-current pulses, carbon on the diamond particle surfaces dissolves in the melted cobalt phase, whereas during the intervals between the pulses the material is rapidly cooled to the assumed densification temperature so that the precipitation (in the form of graphite) of the carbon dissolved in cobalt is hampered. This is why no graphite precipitates occur in the WCCo/diamond composites sintered by the PPS method. Another advantage of the PPS process is that, thanks to the very short duration of the high temperature and its rapid decrease to the stable densification temperature, also the growth of the WC grains is hampered since the precipitation of the tungsten carbide dissolved in cobalt is restricted.

As reported by many investigators, graphitization of diamond is an important problem since it degrades the performance properties of the composites reinforced with diamond particles. Many authors suggest that the presence of graphite particles in a cutting tool results in its short service life since this leads to the diamond particles being thorn out from the matrix during the cutting process.

The possibility of producing WCCo/diamond composites under conditions of thermodynamic instability of diamond is based on the assumptions:

1. At normal pressure, diamond is a metastable phase of carbon whereas at higher temperatures it undergoes transformation into the stable graphite phase.

2. The rate at which the graphitization of diamond proceeds depends on the temperature and time. However in high vacuum (low oxygen partial pressure) and at temperatures up to 1300°C, the graphitization proceeds very slowly: when a diamond powder is heated at a temperature of 1300°C in high vacuum for 1h, only 1-1.5% of the diamond phase is transformed into graphite [19,20].
3. Literature reports concerning graphitization of diamond indicate that by conducting the sintering process under high vacuum at a temperature below 1300°C and for a short time it is possible to restrict the transformation of diamond into graphite or even avoid it.
4. The conditions prevailing in densification by the PPS method permits limiting the rate of diamond graphitization.

Table 1 compares the hardness of various WCCo/diamond composites:

— 30 vol % of diamond particles produced using a mixture of a WC powder (with a particle size of 0.4 μm) added with 6 wt % of Co and a diamond powder (particle size – 16 to 20 μm) produced by PPS [our experiments],

— 30 vol % of diamond particles produced from the mixture of a WC powder (with a particle size of 0.8 μm) added with 6 wt % of Co and a diamond powder (particle size – 40 to 60 μm) produced by PPS [13]

— 20 vol % of diamond particles produced from the mixture of a WC powder (with a particle size of 1.9 μm) added with 10 wt % of Co and a diamond powder (particle size – 8 to 16, 20 to 30 and 40 to 60 μm) produced by SPS [6],

— 20 vol % of diamond particles produced from the mixture of a WC powder (with a particle size of 30 nm) added with 10 wt % of Co and a diamond powder (particle size – 150 to 375 μm) produced by SPS [7],

— 20 vol % of diamond particles produced from the mixture of a WC powder (with a particle size of 1.9 μm) added with 10 wt % of Co and a diamond powder (particle size – 8 to 16μm) produced by PECS [21].

A WCCo/diamond composite with larger diamond particles had higher hardness (at the same diamond vol.%). The hardness of the WCCo/diamond composite with 30 vol % of diamond particles whose size ranged from 40 to 60 μm sintered by PPS was higher by about 2 GPa than that of the WCCo/diamond composite with smaller diamond particles (16-20 μm). The same correlation can be seen in WCCo/diamond composites with 20 vol % of diamond particles sintered by SPS. X.L. Shi et al. [7] give no information about the hardness of the WCCo/diamond composites reported in their paper probably because of the low density of these composites. They found that, in the WC10Co/diamond composite with uncovered diamond particles, the diamond particles were embedded in the matrix only mechanically so that voids occurred between the diamond and the matrix and large graphite grains were present on the diamond particle surfaces.

Method	Temperature [°C]	Load [MPa]	Time [min]	Diamond particle size [μm]	Diamond [vol.%]	Hardness [GPa]	Ref.
PPS	1100	75	5	40-60	30	23	[13]
PPS	1050	100	5	16-20	30	21	This work
SPS	1300	41	3	8-16	20	16.5	[6]
SPS	1300	41	3	20-30	20	17	[6]
SPS	1300	41	3	40-60	20	17.5	[6]
SPS	1280	30	5	150-375	20	-	[7]
PECS	1220	30	5	8-16	20	15.5	[21]

The WCCo/diamond transition layer has a decisive influence on the mechanical properties of the composite. The present experiments have shown that with the decrease of the densification temperature from 1100 to 1050°C the number of the trans-crystalline fractures increases twofold. The XRD examinations of the WCCo/diamond composites, sintered within the temperatures range employed here, revealed no differences in the phase composition of the composite: the only phases identified were diamond, tungsten carbide and cobalt.

CONCLUSIONS

In the proposed technique of producing the WCCo/diamond composites (16-20 μm diamond particles) there is no need for using a specialized High Pressure-High Temperature method. A WC-Co/diamond composite of high density was produced by the PPS method at the densification temperature of 1050°C under the conditions of thermodynamic instability of diamond. The specific conditions prevailing during heating by high-current pulses permitted avoiding graphitization of diamond.

ACKNOWLEDGEMENT

This work was supported by the project No 1070/R/T02/2010/10 from the National Centre for Research and Development.

REFERENCES

[1] S.I. Cha, S.H. Hong, G.H. Ha and B.K. Kim, "Microstructure and Mechanical Properties of Nanocrystalline WC-10Co Cemented Carbides," *Scripta Materialia*, **44** [8] 1535-1539 (2001).

[2] P. Feng, W. Xiong, L. Yu, Y. Zheng, Y. Xia, "Phase evolution and microstructure characteristics of ultrafine Ti(C,N)-based cermet by spark plasma sintering," *International Journal of Refractory Metals & Hard Materials*, **22** [2] 133-138 (2004).

[3] L.H. Zhu, Q.W. Huang, H.F. Zhao, "Preparation of nanocrystalline WC-10Co-0.8VC by spark plasma sintering," *Journal of Materials Science Letters*, **22** [22] 1631-1633 (2003).

[4] J.R. Groza, A. Zavaliangos, "Nanostructured bulk solids by field activated sintering," Advanced Materials Science, **5** 24-33 (2003).

[5] S. Grasso, Ch. Hu, G. Maizza, Y. Sakka, "Spark Plasma Sintering of Diamond Binderless WC Composites," *Journal of the American Ceramic Society*, DOI: 10.1111/j.1551-2916.2011.05009.x.

[6] H. Moriguchi, K. Tsuduki, A. Ikegaya, Y. Miyamoto, and Y. Morisada, "Sintering behavior and properties of diamond/cemented carbides," *International Journal of Refractory Metals & Hard Materials*, **25** [3] 237-243 (2007).

[7] X.L. Shi, G.O. Shao, X.L. Duan, Z. Xiong, H. Yang, "The effect of tungsten buffer layer on the stability of diamond with tungsten carbide–cobalt nanocomposite powder during spark plasma sintering," *Diamond Related Materials*, **15** [10] 1643-1649 (2006).

[8] M. Rosinski, A. Michalski, "Nanocrystalline NiAl-TiC Composites Sintered by the pulse Plasma Method," *Solid State Phenomena*, **114** 233 (2006).

[9] A. Michalski, M. Rosinski, D. Siemiaszko, J. Jaroszewicz, K. J. Kurzydłowski, "Pulse plasma sintering of nano-crystalline Cu powder," *Solid State Phenomena*, **114** 239 (2006).

[10] A. Szymanska, D. Oleszak, A. Grabias, M. Rosinski, K. Sikorski, J. Kazior, A. Michalski, K. J. Kurzydlowski, "Phase transformations in ball milled AISI 316L stainless steel powders and the microstructure of the steel obtained by its sintering," *Rev. Adv. Mater. Sic.*, **8** 143 (2004).

[11] M. Rosinski, A. Michalski, M. Płocinska, J. Szawlowski, "WC/Ti composite material enriched with cBN particles produced by pulse plasma sintering (PPS)," *Key Engineering Materials*, **484** 130 (2011).

[12] M. Rosinski, A. Michalski, "WCCo/cBN composites produced by pulse plasma sintering method", *Journal of Materials Science*, available online 10.1007/s10853-012-6532-x

[13] A. Michalski, M. Rosinski, "Sintering diamond/cemented carbides by the pulse plasma sintering method," *Journal of the American Ceramic Society*, **91** [11] 3560-3565 (2008).

[14] [31] M. Rosinski, L. Ciupinski, J. Grzonka, A. Michalski and K.J. Kurzydlowski, "Synthesis and characterization of the diamond/copper composites obtained by the pulse plasma sintering (PPS) method," *Diamond & Related Materials*, **27–28** 29 (2012).

[15] T. Wejrzanowski "Computer Assisted Quantitative Description of the Functionally Graded Materials," M.Sc. thesis, Warsaw University of Technology, Warsaw, Poland (2000).

[16] T. Wejrzanowski, W.L. Spychalski, K. Rozniatowski, K.J. Kurzydłowski, "Image based analysis of complex microstructures of Engineering materials," *International Journal of Applied Mathematics and Computer Science - Applied Image Processing*, **18** [1] 33–39 (2008).

[17] V. Martínez, J. Echeberria, "Hot isostatic pressing of cubic boron nitride-tungsten carbide/cobalt (cBN-WC/Co) composites: Effect of cBN particle size and some processing parameters on their microstructure and properties," *Journal of the American Ceramic Society*, **90** [2] 415–424 (2007).

[18] G.S. Upadhyaya, "Cemented Tungsten Carbides: Production, Properties & Testing," Noyes Pablications, Fairfield, NJ, USA, (1998).

[19] K.C. Uspenskaya, U.N. Tormashev and D.V. Fedoceev, "Oxidization and graphitization of diamond in condition of low pressure atmosphere," *Journal of Physics and Chemistry*, **56** [2] 495-496 (1982) (in Russian).

[20] W.Z. Shao, V.V. Ivanov, L. Zhen, Y.S. Cui and Y. Wang, "A study on graphitization of diamond in copper–diamond composite materials," *Materials Letters*, **58** 146-149 (2003).

[21] Y. Miyamoto, T. Kashiwagi, K. Hirota, O. Yamaguchi, H. Moriguchi, K. Tsuduki, A. Ikegaya, "Fabrication of New Cemented Carbide Containing Diamond Coated with Nanometer-Sized SiC Particles," *Journal of the American Ceramic Society*, **86** [1] 73-76 (2003).

SURFACE MORPHOLOGY OF YSZ THIN FILMS DEPOSITED FROM A PRECURSOR SOLUTION UNDER THE ELECTRICAL FIELDS

Atsushi Saiki, Kento Hamada and Takashi Hashizume
Field of Nano and Functional Material Science, Graduate School of Science and Engineering for Research, University of Toyama
Toyama-shi, Toyama, 930-8555, Japan

ABSTRACT

In this study, surface morphology control and patterning has been carried out during deposition of thin films from a precursor solution by applying electrochemical method and electrical filed at the same time. The precursor was an aqueous solution of zirconyl nitrate, $Y(NO_3)_3$ -$6H_2O$, and NH_3(aq). The thin film was deposited on the glass substrate of the minus electrode side. By applying the electrical field of 2.3-3.0 V, the $Zr(Y)(OH)_4$ thin film was effectively deposited on glass substrates at room temperature. The as-deposited film was amorphous, and a crystalline phase of YSZ can be obtained after annealing at 723 K for 6 h in air. In order to establish morphology control and patterning, another pulse bias was applied to the electrical field along the perpendicular direction from the film deposition direction. The pulse type, frequency, amplitude, duty ratio were changed and the effect to the surface morphology was investigated. At only in limited condition, the striped patterns of YSZ films due to the frequency of the applied pulsed electrical field were observed.

INTRODUCTION

Zirconia is critical for electrical property and chemical, thermal, mechanical stability[1, 2], and its thin films have attracted much attention for such applications as the miniaturization of solid oxide fuel cells (SOFCs)[3, 4], optical coatings[5], thermal-shield or corrosion-resistant coatings[6], oxygen sensors[7], and buffer layers of electric devices[8]. Thin and dense films are needed for these applications. In addition patterning techniques are also searched. To improve such properties many YSZ films are fabricated by PVD or CVD methods including pulsed laser deposition (PLD), magnetron sputtering[9], metal organic chemical vapor deposition (MOCVD)[10] and electron-beam physical vapor deposition (EB-PVD)[11]. On the other hand liquid phase deposition (LPD) methods such as electrochemical deposition (ECD)[12, 13, 14], electrophoresis deposition (EPD)[15], and mist deposition[16], and spin coating [17] have also been used for relatively thick films. One fabrication method for zirconia thin films using relatively environment friendly materials is to use aqueous solutions as raw materials. By using the decomposition reaction of an aqueous peroxozirconium complex solution with a zirconium oxynitrate dihydrate ($ZrO(NO_3)_2$-$2H_2O$) as a Zr source[16, 18], pure amorphous ZrO_2 precursor thin film was deposited at room temperature. In our previous work, we reported that YSZ thin film was effectively deposited on the minus electrode gap side of the glass substrate from a zirconium oxynitrate dihydrate solution by applying a constant electrical field[14, 19].

In this study we investigated the effect of another pulse bias applied to the electrical field in the aqueous solution along the perpendicular direction from the film deposition direction.

EXPERIMENTAL PROCEDURE

Zirconyl nitrate and Yttrium nitrate hexahydrate ($Y(NO_3)_3$-$6H_2O$, 99.99%.) were selected as the Zr and Y sources. The composition and concentration of the precursor aqueous solution were adjusted to 92 mol%ZrO_2-8mol%$YO_{1.5}$ and 0.05 mol/dm^3. Appropriate quantities (0-0.5 vol%) of an ammonia aqueous solution (NH_3(aq), 28 %) were added to adjust the acidity and to promote a radical exchange reaction between the $ZrO(NO_3)_2$ and NH_4OH to produce an intermediate product of $ZrO(OH)_2$. After the mixture was stirred for 1 h in an ice bath, a homogeneous colorless and transparent solution was obtained. Glass substrates (9.0 x 9.0 x 0.15 mm^3) were cleaned ultrasonically in an ethanol and glass cleaner (Semico clean 56) and irradiated by UV light for 20 min to improve wettability. Fig. 1 is a schematic diagram of the cell for thin film deposition. The glass substrate was placed above the minus electrode with a gap distance of 96 μm. The YSZ thin film was deposited on the minus electrode side of the glass substrate by applying a constant electrical field of 3.0 V for 1200 s. And at same time another pulse bias was applied to the electrical field along the perpendicular direction from the film deposition direction. In this study, a triangle square type pulsed electrical field was used, whose duty ratio, amplitude, and bias voltage (DC offset) were fixed to 50:50, 2.50 V, and 0 V, respectively as shown in Fig. 2, and its frequency was changed from 0 to 10 Hz. Since as-deposited films were smooth and almost transparent but amorphous, heat treatment was performed at 773 K for 6 h afterwards to obtain crystallized YSZ.

The crystallization of the films was investigated using an x-ray diffractometer (XRD, 40 kV, 30 mA, CuKa, XRD-6100, Shimadzu). Their chemical composition was analyzed by an x-ray fluorescence spectrometer (EAGLE μ -Probe, EDAX and EPMA JXA-8230, JEOL). The film morphology was observed by scanning electron microscope (SEM, TM-1000, HITACHI) and optical microscope (BHSM-313MB, OLYMPUS).

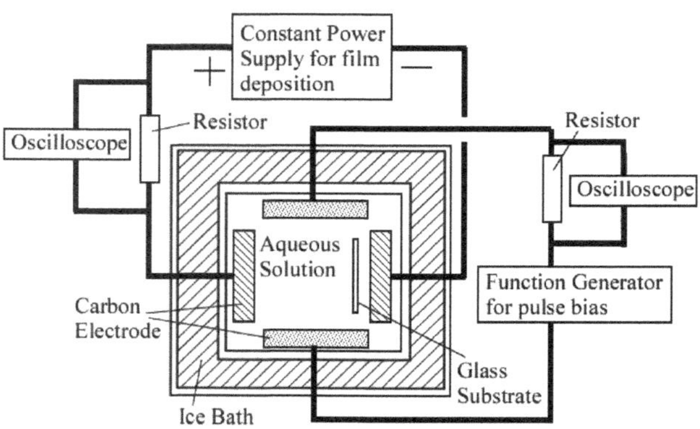

Fig.1 Schematic diagram of the cell for thin film deposition by the ECD method with the pulsed bias assist

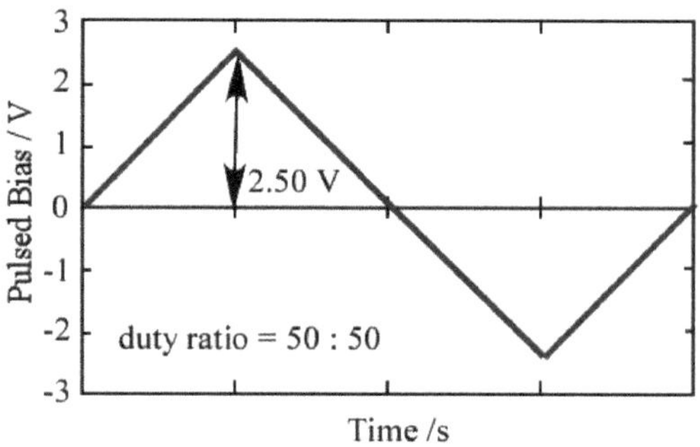

Fig.2 Applied pulsed bias to the electrical field in the aqueous solution along the perpendicular direction from the film deposition direction.

RESULT AND DISCUSSION

Film Deposition

During the film deposition, a constant voltage of 3.0 V was applied between the electrodes for film deposition. The current value decreased steeply in the first several minutes just after the voltage was applied as shown in Fig. 3 (a), which indicated the initial ion movement based on the applied electrical field and the formation of an electric double layer near the electrode. After that the current almost became stable and the film began to precipitate. On the other hand, when another pulse bias was applied to the electrical field in the aqueous solution along the perpendicular direction from the film deposition direction as shown in Fig. 3 (b) – (f), the current value oscillated as the frequency of the applied pulse bias, except Fig 3 (e) 5 Hz.

As-deposited films were smooth and transparent but amorphous, and no obvious x-ray diffraction peaks were detected. For that reason samples were heat-treated at 773 K for 6 h afterwards to obtain crystallized YSZ. Well crystallized films with smooth surfaces and few cracks were formed when the applied voltage was 3 V and the frequency of the applied pulse bias was 0 Hz, as shown in Fig.4 (a). When voltage exceeding 3.5 V was applied, the deposition rate increased due to the electrolysis of water. When the amount of added NH_3(aq) was larger in the range of 0 to 0.5 vol%, the thicker films were also deposited in the same deposition time. However when the as-deposition film thickness exceeded about 2 μm or the deposition rate was too fast, it was not favorable because many cracks tended to develop during the heat treatment.

In this method, since a radical exchange and precipitation reaction occurred with the assistance of the electric field, they were influenced by the applied pulse bias voltage along the perpendicular direction from the film deposition direction. The stability of the $ZrO(NO_3)_2$, $ZrO(OH)_2$, and the deposition of the film was also affected by the subtle acidity change near the substrate. As the result, YSZ films deposited only on the specific position like the pinstripes as shown in Fig. 4 (b) – (d). The direction of the pinstripes was almost normal to the direction of applied pulse bias. Fig.5 was the line analysis of (a) Zr and (b) Y along normal to the pinstripe direction by EPMA. Between pinstripes only little amount of YSZ film deposited. And at the same time it was confirmed that the concentrations of Y in YSZ film on pinstripes and that of the aqueous solution were also constant. The distance between the pinstripes changed with the frequency of the pulse bias as shown in Fig.6.

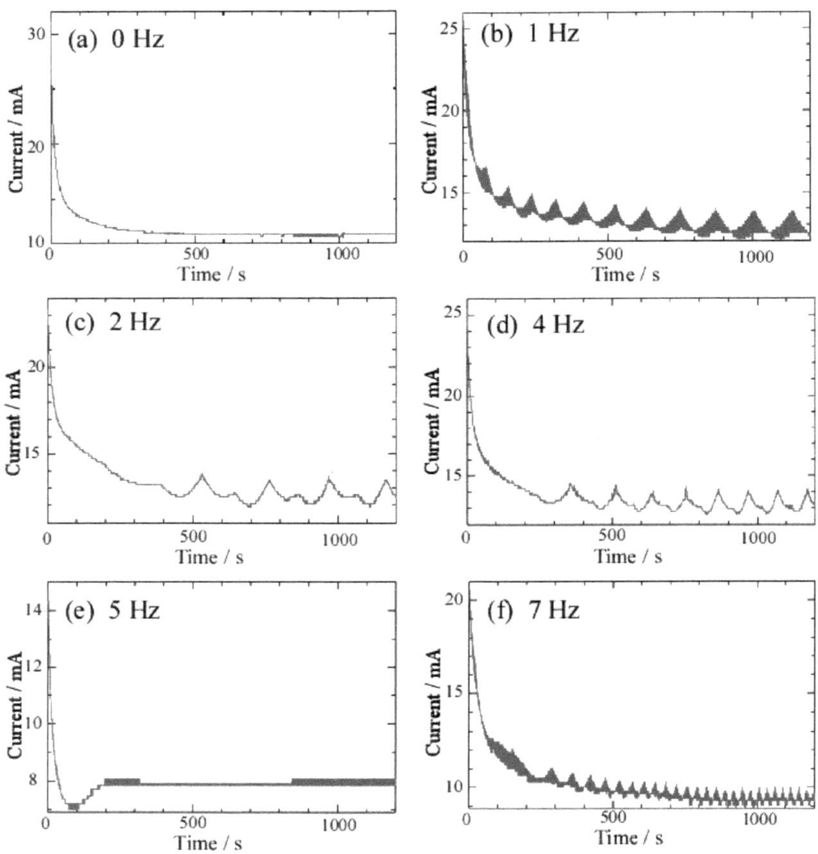

Fig.3 Influence of the frequency of the applied pulse bias on deposition current at thin films deposition

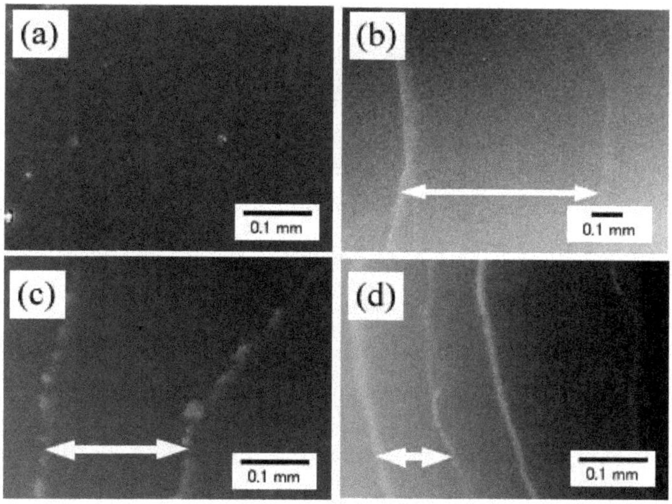

Fig.4 SEM micrographs of YSZ thin films fabricated under the applied pulse bias to the electrical field ; (a) 0 Hz, (b) 1 Hz, (c) 5 Hz, (d) 10 Hz. Arrows indicate the direction of applied pulse bias.

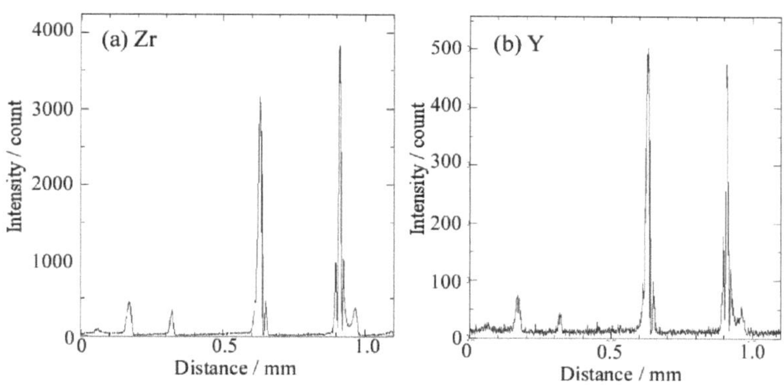

Fig.5 Line analysis of (a) Zr and (b) Y by EPMA. YSZ thin film was fabricated under the applied pulse bias of 5 Hz. Analyzed direction accorded with pulse applied direction.

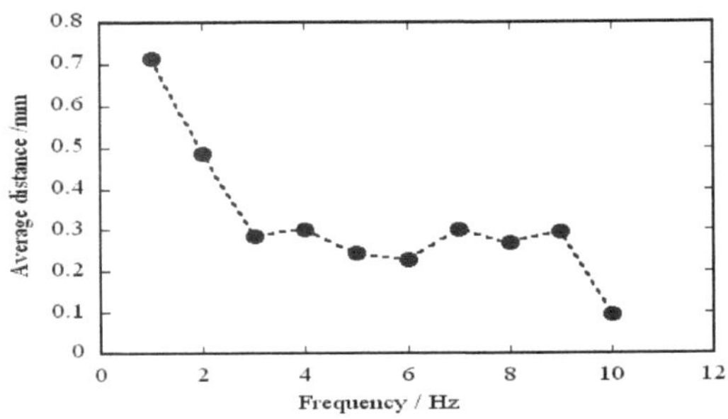

Fig.6 Average distance between the pinstripes of the YSZ film deposited in the aqueous solution under the applied pulse bias to the electrical field. Dashed line shown as a visual aid.

The following was a schematic ion movement during deposition. At the surface of the minus electrode, hydrogen ions gather around the electrode and an electric double layer is formed[20]. Then mutual repulsion between the ions and the hydroxide ion concentration also rose locally near the glass substrate. Therefore the film was deposited on the glass substrate surface of the minus electrode side. In this study the pulsed bias was also applied, and it made the stationary wave in the solution near the film deposition place, and only at the section it was possible to deposit the film because the pulse bias did not affect at that position. The wavelength of the stationary wave changed as the applied pulse bias, therefore the distance of pinstripe also changed. On of this patterning method was unique compared to the conventional EPD method.

Influence of pulsed electrical field

When a pulsed electrical field was applied, the current value oscillated as the frequency of the applied pulse bias as shown Fig.3. In the figure the oscillation modes of (c) 2 Hz and (d) 4 Hz, or (b) 1 Hz and (f) 7 Hz were resemble each other. In case of (c), the average cycle of the current oscillation was 215 sec and that of (d) was 130 sec. On the other hand the average distance between pinstripe of (c) was 0.49 mm and that of (d) was 0.30 mm. From these values, next relationship was conducted

$$\frac{0.30}{0.49} \approx \frac{130}{215} \approx 0.61 \approx \text{ constant} \qquad (1)$$

In case of (b) 1 Hz and (e) 7 Hz similar relationship was also conducted

$$\frac{0.30}{0.71} \approx \frac{38}{90} \approx 0.42 \approx \text{ constant} \qquad (2)$$

The distance between the pinstripes changed with the frequency of the pulse bias, and it was confirmed that when the oscillation modes resembled each other, there existed the correlation between the distances of the pinstripes and the cycle of the current value oscillation.

CONCLUSION

In our present study, YSZ thin films were grown on glass substrates in an aqueous complex solution with the assistance of a constant electrical voltage for film deposition and pulsed electrical fields for morphology control at the same time. $ZrO(NO_3)_2$-$2H_2O$ and $Y(NO_3)_3$-$6H_2O$ were used as Zr and Y sources. YSZ precursor thin film was deposited on the minus electrode side of the glass substrate that was placed above the carbon minus electrode with a gap distance of 96 μm by applying voltage of 3.0 V for 1200 s. As-deposited films were smooth and transparent but amorphous. After heat treatment at 773 K for 3 h, crystallized YSZ films were obtained. When the pulsed electrical field was applied normal to the direction of the film deposition, the YSZ films deposited only on the specific position like the pinstripes. The distance between the pinstripes changed with the frequency of the pulse bias, and it was confirmed that there existed the correlation between the distances of the pinstripes and the cycle of the current value oscillation.

REFERENCES

[1] O. Vasylkiv, Y. Sakka, Y. Maeda, V. V. Skorokhod, J. Am. Ceram. Soc., **88**(2005) 639-644

[2] J. Eichler, U. Eisele, J. Rodel, J. Am. Ceram. Soc., **87**(2004) 1401-1403

[3] K. Rodrigo, J. Knudsen, N. Pryds, J. Schou, S. Linderoth, Appl. Surface Scie., **254** (2007) 1338–1342

[4] D. Marrero-López, J.C. Ruiz-Morales, J. Peña-Martínez, J.Canales-Vázquez, P. Núñez, J. Solid State Chem., **181** (2008) 685-692

[5] H. Wendel, H. Holzschuh, H. Suhr, G. Erker. S. Dehnicke, M. Mena, Mod. Phys. Lett. B **4**, 1215-1217 (1990)

[6] G. R. Dickinson, C. Petorak, K. Bowman, R. W. Trice, J. Am. Ceram. Soc., **88**(2005) 2202-2208

[7] A. Bastianini, G. A. Battiston, R. Gerbasi, M. Porchia, S. Daolio, J. Phys **5** (1995) 525-532.

[8] N. Wakiya, K. Shinozaki, N. Mizutani, Key Engineering Materials, **269** (2004) 245-248

[9] P. Briois, F. Lapostolle, V. Demange, E. Djurado, A.Billard, Surface and Coatings Tech., **201**(2007) 6012-6018

[10] Yinzhu Jiang, Jianfeng Gao, Mingfei Liu, Yanyan Wang, Guangyao Meng, Solid State Ionics, **177**(2007) 3405-3410

[11] I.M. Ochando, D. Cáceres, J. García-López, R. Escobar-Galindo, R.J. Jiménez-Rioboó, C. Prieto, Vacuum, **81**(2007) 1457-1461

[12] J.Kim, Y.Park, D.Jin-Sung, S.Moon, K.Bong-Lee, S.In Hong, Refractory Metals & Hard Materials, **27**(2009)985-990

[13] T.Talebi, B.Raissi, M.Haji, A.Maghsoudipour, Hydrogen energy, **35**(2010)9405-9410

[14] A.Saiki, H.Uno, S.Ui. Y.Fujisawa, T.Hashizume, K. Terayama, Ceram.Trans., **195**(2006)77-83

[15] Li Jia, Zhe Lü, Xiqiang Huang, Zhiguo Liu, Kongfa Chen, Xueqing Sha, Guoqing Li, Wenhui Su, J. Alloys and Compounds, **424**(2006) 299-303

[16] A. Saiki, Y. Fujisawa, T. Hashizume, K. Terayama, Ceram.Trans., **195**(2006)115-121

[17] Kongfa Chen, Zhe Lü, Na Ai, Xiqiang Huang, Yaohui Zhang, Xiaodong Ge, Xianshuang Xin, Xiangjun Chen, Wenhui Su, Solid State Ionics, **177**(2007) 3455-3460

[18] Y. Gao, Y. Masuda, H. Ohta, K. Koumoto, Chem. Mater. **16**(2004) 2615-2622

[19] A.Saiki, T.Hashizume, K.Terayama, Materials Transaction, **51**(2010)261-264

[20] A.Saiki, C.Kawai, T.Hashizume, K.Terayama, Materials Science and Engineering, **18**(2011)032011 doi:10.1088/1757-899X/18/032011

Author Index